Urban Sustainability

The Urban Sustainability Book Series is a valuable resource for sustainability and urban-related education and research. It offers an inter-disciplinary platform covering all four areas of practice, policy, education, research, and their nexus. The publications in this series are related to critical areas of sustainability, urban studies, planning, and urban geography.

This book series aims to put together cutting-edge research findings linked to the overarching field of urban sustainability. The scope and nature of the topic are broad and interdisciplinary and bring together various associated disciplines from sustainable development, environmental sciences, urbanism, etc. With many advanced research findings in the field, there is a need to put together various discussions and contributions on specific sustainability fields, covering a good range of topics on sustainable development, sustainable urbanism, and urban sustainability. Despite the broad range of issues, we note the importance of practical and policy-oriented directions, extending the literature and directions and pathways towards achieving urban sustainability.

The series will appeal to urbanists, geographers, planners, engineers, architects, governmental authorities, policymakers, researchers of all levels, and to all of those interested in a wide-ranging overview of urban sustainability and its associated fields. The series includes monographs and edited volumes, covering a range of topics under the urban sustainability topic, which can also be used for teaching materials.

Ali Cheshmehzangi · Sara Alidoust ·
Wendy Y. Chen · Richard Fuller
Editors

Nature in Cities, Nurturing Cities

Springer

Editors
Ali Cheshmehzangi
Architecture, Design and Planning
University of Queensland
Brisbane, QLD, Australia

Sara Alidoust
Architecture, Design and Planning
University of Queensland
Brisbane, QLD, Australia

Wendy Y. Chen
Department of Geography
University of Hong Kong
Hong Kong, Hong Kong

Richard Fuller
School of the Environment
University of Queensland
Brisbane, QLD, Australia

ISSN 2731-6483 ISSN 2731-6491 (electronic)
Urban Sustainability
ISBN 978-981-95-5486-7 ISBN 978-981-95-5487-4 (eBook)
https://doi.org/10.1007/978-981-95-5487-4

This Springer imprint is published by the registered company Springer Nature Singapore Pte Ltd.
The registered company address is: 152 Beach Road, #21-01/04 Gateway East, Singapore 189721, Singapore

We collectively dedicate this book to the passionate defenders and champions of nature, whose tireless work continues to inspire change.

This goes to good and genuine human beings like David Attenborough. Through his unwavering efforts to educate people across cultures and continents, David has shown that understanding nature is not just a scientific pursuit but a moral responsibility, one that binds humanity to the planet we call home. There are so many good people like him; influential leaders and activists like Nonhle Mbuthuma, Mahlagha Mallah, Ali Yachkaschi, Jane Goodall, Rachel Carson, Manuel Esteban Paez Terán, Liang Congjie, Daniel Piedrahíta, and many more.

Preface

Nature does not hurry, yet everything is accomplished.

—Lao Tzu

We are at a pivotal point in the development of urban civilization. For over a century, cities have been constructed as testaments to human ingenuity, combining steel, concrete, glass, and speed to create ambitious landscapes. However, beneath the noise of progress, something crucial has been overlooked. We have not viewed nature as the living basis that enables our accomplishments, but rather as ornamentation, a beautiful setting for them. Unquestionably, that attitude has cost us: overheated streets, fractured communities, exhausted ecosystems, and a public health crisis measured not only in disease but in disconnection and decline.

This book is a call to arms, not to abandon cities but to transform them. It makes the case that reintegrating urban life with the natural systems that support us is essential to its future. Nature must no longer be considered a luxury or a mark of wealth. It must be acknowledged as key infrastructure, just as vital to a city's operation as digital networks, electrical grids, and water pipes. Wetlands and trees are not decorative. They act as air filters, water managers, climate regulators, and biodiversity havens. Gardens and parks are more than just places to relax. They serve as both social cohesion labs and mental health pillars. Coastal wetlands, green roofs, and river corridors are not aesthetic gestures. They serve as our first and most effective line of defense against the chaos brought on by global warming.

But the goal here is to restore meaning, not just solve issues. In urban settings, nature has the capacity to reawaken a stronger sense of identity, anchor us, and serve as a reminder that we are a part of something bigger and more ancient than ourselves. More than just food is produced when communities work together to create gardens; trust and a common goal are also fostered. Children who play under the shade of urban trees gain a living connection to the planet's history in addition to some shade. Design interventions and solutions that value ecosystem rhythms produce spaces that encourage care rather than consumption.

What follows is a transformational plan that is based on research, shaped by data, and validated by practice, rather than an idealistic manifesto. This book demonstrates how incorporating nature back into urban areas may improve human well-being, increase resiliency to climate extremes, and create areas that promote community and identity. It urges citizens, designers, planners, and legislators to reject the erroneous decision to choose between ecological integrity and urban growth. The future we require is a combination of both, not just one.

There is no time left for hesitation. The choices we make today will determine whether our cities become hubs of vibrant life or symbols of shortsightedness as the climate problem worsens and urban populations grow. Nature is an active ally that awaits an invitation to return, not a helpless victim. It contains the knowledge, fortitude, and capacity for regeneration that we require to face the coming century.

Nature in Cities, Nurturing Cities makes the case for having the guts to rethink our urban destiny in a way that prioritizes life. It serves as a reminder that a living breathing city where people and nature coexist peacefully is the best legacy we can create, not a skyline.

Brisbane, Australia — Ali Cheshmehzangi
Brisbane, Australia — Sara Alidoust
Hong Kong, Hong Kong — Wendy Y. Chen
Brisbane, Australia — Richard Fuller

Acknowledgements We acknowledge everyone's contributions to this exciting book project. We extend our gratitude to all the authors, co-authors, chapter contributors, and collaborators who have shared their expertise, passion, and creativity with us. This sentiment applies to our earlier and forthcoming work as well, where we deeply value the dedication and vision of all contributors who helped create this book. Such collaborative efforts foster new dialogues, networking opportunities, and a valuable research exchange platform for scholars worldwide.

About This Book

The transforming potential of nature as a key factor in creating urban settings that are healthier, more resilient, and more inclusive is examined in ***Nature in Cities, Nurturing Cities***. This book reimagines the relationship between urban development and ecological integration by combining multidisciplinary research, real-world case studies, and innovative techniques within the expanding area of urban sustainability. This volume presents nature as the fundamental infrastructure for sustainable urban living rather than as an aesthetic afterthought in the face of cities' mounting problems, which range from biodiversity loss and climate change to public health disparities.

A central pillar of this book is its focus on three interconnected dimensions that define the next frontier of urban nature integration. The first, *Nature and Human Wellbeing*, examines the direct effects that having access to green and blue areas has on mental recovery, physical health, and daily quality of life. Exposure to biodiverse landscapes has been shown to reduce stress, improve cognitive function, lessen the risk of chronic diseases, and increase overall pleasure, according to research from environmental psychology, public health, and urban planning. The objective here is to incorporate nature into the rhythms of everyday life, from street trees and edible gardens to green roofs and living walls, rather than just sprinkling more parks around a city. In addition to being aesthetically pleasing, these components establish an ongoing system of healing spaces that transform how people view and engage with the city.

The second theme, *Climate Resilience and Urban Cooling*, discusses how nature plays a crucial part in shielding urban areas from increasingly severe climate conditions. Urban heat islands, which occur when densely populated areas are several degrees warmer than the surrounding landscape, are currently one of the most pressing issues facing infrastructure and public health. Urban trees, wetlands, and water-sensitive architecture are examples of nature-based solutions that can reduce surface and ambient temperatures, manage stormwater, sequester carbon, and improve air quality. These measures are essential for adaptation in addition to mitigation because they enable cities to retain their liveability in the face of future climate scenarios, absorb shocks, and recover more quickly. This section outlines design strategies and policy choices to elevate ecological performance from a secondary benefit to a primary metric of urban development.

Social Belonging and Placemaking, the third theme, examines how natural settings help people develop stronger bonds with their surroundings. Well-planned green areas can serve as civic gathering places, fostering social cohesiveness, cultural identity, and collective memory. The presence of nature can heal divisions and fortify the social fabric, as seen in anything from community gardens that encourage local care to restored riverfronts that reunite locals with their common heritage. Ecologically based placemaking turns public areas into democratic, inclusive settings that foster a sense of community and encourage environmental stewardship.

Altogether, ***Nature in Cities*, *Nurturing Cities*** makes a compelling case for incorporating nature into the urban fabric in a deliberate, inclusive, and resilient manner. It makes the case that nature should no longer be viewed as an add-on but rather as an essential component of urban life, supporting social vitality, climate stability, and public health. The book, as the first of three volumes on the topic of "nature in cities" topic, shows how nature may simultaneously enhance environmental performance, build community resilience, and improve the lived experience of cities by redefining ecological assets as vital infrastructure. In order to include biodiversity, human well-being, and social connection into all facets of urban decision-making, from neighborhood-scale initiatives to metropolitan strategies, it advocates for a paradigm shift in urban planning and design.

Key Features and Benefits:

- **A Timely book focused on reimagining of urban futures through nature.**
- **Provides a comprehensive collection grounded in evidence, enriched by global case studies.**
- **Serves as a unique toolkit for innovation toward nature-based urbanism.**

Contents

Editors and Contributors

About the Editors

Ali Cheshmehzangi is Head of the School of Architecture, Design and Planning and a Professor at The University of Queensland (UQ), Brisbane, Australia. He is among the top 20 global scholars in the urban sustainability research area. He is the Editor-in-Chief of the Urban Sustainability Book Series with Springer Nature. With a career spanning over two decades, he has made significant contributions to the academic and professional communities, with a focus on sustainable and environmentally conscious design. So far, Ali has published over 500 journal papers, articles, conference papers, book chapters, and reports. He has published more than 50 academic books, some of which have received awards at the international, national, provincial, and municipal levels. He also has received international awards and recognition for his research on urban resilience studies and sustainability research, as well as the 2018 Vice-Chancellor's award for his impactful contribution to higher education.

Sara Alidoust is a Senior Lecturer in Urban Planning at The School of Architecture, Design and Planning at The University of Queensland (UQ), Brisbane, Australia. Her research centers on the intersections of Urban Planning and Public Health, with a strong emphasis on interdisciplinary collaboration and practical application. Sara delivers transdisciplinary solutions to some of the most compelling challenges of our time, 'housing equity' and 'healthy cities'. Her work stands out through the application of Systems Thinking to complex planning issues and she explores the connections between planning and the physical, mental, and social health of individuals and communities. Her primary focus is on developing resilient cities, capable of maintaining their liveability over time, especially amidst disruptive events and shocks.

Wendy Y. Chen is a professor in Department of Geography at The University of Hong Kong. She serves as the Editor-in-Chief for Urban Forestry & Urban Greening, a top international journal in the field of urban greening design and management. Her research agenda has focused on key knowledge gaps in evaluating and modelling urban green-blue spaces (GBS) that are reserved, modified, and deployed in urban/peri-urban areas. She is amongst a small group of scholars who have begun to experiment and validate the utilisation of classical non-market approaches in China's transitional context and facilitate the quantitative assessment of GBS' contribution to the quality of life. An additional strand of her research is dedicated to investigating GBS dynamics, pertaining to how and why GBS as public environmental goods are physically transformed, economically incorporated, and socially mobilized. Her research helps to integrate socioeconomic dimensions to enrich GBS scholarship, and rethink theories pertaining to GBS demand and supply in the context of extensive urbanization in developing countries and re-urbanization in developed world. She is ranked as a top 1% scholar worldwide by Clarivate Analytics.

Richard Fuller is a leading expert in urban ecology research. He is a Professor at School of the Environment at The University of Queensland (UQ), Brisbane, Australia. He works on pure and applied topics in biodiversity and conservation. Much of his work is interdisciplinary, focusing on the interactions between people and nature, how these can be enhanced, and how these relationships can be shaped to converge on coherent solutions to the biodiversity crisis. His current research topics include the ecology of urban environments, understanding what drives some people to show stronger environmental concern than others, and strategies for designing efficient conservation plans. He enjoys working closely with all his wonderful colleagues in the Centre for Biodiversity and Conservation Science.

Contributors

Sara Alidoust School of Architecture, Design and Planning, The University of Queensland, Brisbane, QLD, Australia

Judy Bush Faculty of Architecture, Building and Planning, The University of Melbourne, Melbourne, Australia

Xavier Cadorel Faculty of Architecture, Building and Planning, The University of Melbourne, Melbourne, Australia

Wendy Y. Chen Department of Geography, The University of Hong Kong, Pok Fu Lam, Hong Kong

Ali Cheshmehzangi School of Architecture, Design and Planning, The University of Queensland, Brisbane, QLD, Australia;
School of Architecture and Urban Planning, Hunan University, Changsha, China;

College of Architecture and Urban Planning (CAUP), Tongji University, Shanghai, China

Pui Kwan Cheung School of Agriculture, Food and Ecosystem Sciences, The University of Melbourne, Burnley, VIC, Australia

Thami Croeser ICON Science, Centre for Urban Research, School of Global, Urban and Social Studies, RMIT University, Melbourne, VIC, Australia

Marie C. Dade School of Agriculture, Food and Ecosystem Sciences, The University of Melbourne, Burnley, VIC, Australia

Kenneth B. H. Er National Parks Board, Singapore, Singapore

Ziyu Fan Department of Urban Planning and Design, Xi'an Jiaotong-Liverpool University, Suzhou, China

Reza Farhadi Department of Human Geography and Planning, Faculty of Geography, University of Tehran, Tehran, Iran

Richard Fuller School of the Environment, The University of Queensland, Brisbane, Australia;
Centre for Biodiversity and Conservation Science, The University of Queensland, Brisbane, QLD, Australia

Rafia Gulzar Department of Architecture, University of Management and Technology, Lahore, Pakistan;
School of Architecture, University of Lahore, Lahore, Pakistan

Saima Gulzar Department of Architecture, University of Management and Technology, Lahore, Pakistan

Jacinta Humphrey ICON Science, Centre for Urban Research, School of Global, Urban and Social Studies, RMIT University, Melbourne, VIC, Australia

C. Y. Jim Department of Social Sciences and Policy Studies (SSPS), The Education University of Hong Kong, Hong Kong, China

John Kandulu College of Business, Government and Law, Flinders University, Adelaide, SA, Australia

Theresa W. L. Lam School of Agriculture, Food and Ecosystem Sciences, The University of Melbourne, Burnley, VIC, Australia

Yanhui Lei Department of Urban Planning and Design, Xi'an Jiaotong-Liverpool University, Suzhou, China

Stephen J. Livesley School of Agriculture, Food and Ecosystem Sciences, The University of Melbourne, Burnley, VIC, Australia

Melanie Lowe Centre for Urban Research, School of Global, Urban and Social Studies, RMIT University, Melbourne, VIC, Australia

Iderlina Mateo-Babiano Faculty of Architecture, Building and Planning, The University of Melbourne, Melbourne, Australia

Fatemeh Mirzaeipour Meybodi Department of Landscape Architecture, Shahid Beheshti University, Tehran, Iran

Saad Mujahid Department of Architecture, University of Management and Technology, Lahore, Pakistan

Saeedeh Nasehi Department of Environmental Planning, Management, and Education, Faculty of Environment, University of Tehran, Tehran, Iran

Maryam Noroozi Department of Environmental Design Engineering, University of Tehran, Tehran, Iran

Rachel R. Y. Oh Helmholtz Centre for Environmental Research, UFZ, German Centre for Integrative Biodiversity Research (iDiv) Halle-Jena-Leipzig), Leipzig, Germany;
National University of Singapore, Singapore, Singapore

Nayanesh Pattnaik Strategic Landscape Planning and Management, School of Life Sciences, Technical University of Munich, Munich, Germany

Stephan Pauleit Strategic Landscape Planning and Management, School of Life Sciences, Technical University of Munich, Munich, Germany

Mohammad A. Rahman School of Agriculture, Food and Ecosystem Sciences, The University of Melbourne, Burnley, VIC, Australia

Hadi RezaeiRad Department of Urbanism, Faculty of Art and Architecture, Bu-Ali Sina University, Hamedan, Iran

Amirhosein Shabani Department of Urban Planning, Na.C, Islamic Azad University, Najafabad, Iran;
Advancement in Architecture and Urban Planning Research Center, Na.C, Islamic Azad University, Najafabad, Iran

Saba Sheikhi Department of Urban and Regional Planning, Faculty of Art, Tarbiat Modares University, Tehran, Iran

Angelia Sia National Parks Board, Singapore, Singapore

Veronica Soebarto School of Architecture and Civil Engineering, The University of Adelaide, Adelaide, SA, Australia

Hadi Soltanifard Department of Environment, Hakim Sabzevari University, Sabzevar, Iran

Shima Taheri Department of Architecture, Isfahan (Khorasgan) Branch, Islamic Azad University, Isfahan, Iran

Puay Yok Tan National Parks Board, Singapore, Singapore;
National University of Singapore, Singapore, Singapore

Yonatal Tefera Centre for Health in All Policies Research Translation, South Australian Health and Medical Research Institute, Adelaide, SA, Australia

Tracy Washington School of Architecture and Built Environment, Queensland University of Technology, Brisbane, Australia

Carmel Williams Centre for Health in All Policies Research Translation, South Australian Health and Medical Research Institute, Adelaide, SA, Australia; School of Public Health, The University of Adelaide, Adelaide, SA, Australia

Jason Wright National Parks Board, Singapore, Singapore

Sifan Yu Department of Urban Planning and Design, Xi'an Jiaotong-Liverpool University, Suzhou, China

Parisa Ziaesaeidi School of Engineering, Design and Built Environment, Western Sydney University, Sydney, Australia

Nature in Cities: Reimagining Urban Futures Through Ecology and Care

Ali Cheshmehzangi, Sara Alidoust, Wendy Y. Chen, and Richard Fuller

Abstract This introductory chapter introduces a redefined vision of urbanism that transcends the outdated divide between built form and natural systems. This first volume of a trilogy on *Nature in Cities* argues that cities cannot remain resilient, equitable, or liveable without fully embedding ecological principles into their identity, planning, and governance. Drawing from multidisciplinary perspectives, it frames nature not as an aesthetic addition but as critical infrastructure for health, climate resilience, inclusivity, and placemaking. The chapter begins by outlining the historical narrative of separation between cities and nature and its enduring consequences for urban vulnerability. It then establishes the transformative potential of ecological integration, showing how access to biodiverse landscapes supports public health and well-being, mitigates climate risks, and fosters social belonging. The book's purpose is to clarify both intellectual aims and practical goals: bridging theory and practice, connecting disciplines, and cultivating a philosophy of care that reimagines the urban condition itself. Positioned within the global sustainability discourse, this work sets the stage for a long-term narrative of nurturing cities, places where ecological care becomes central to survival, justice, and dignity in our time.

A. Cheshmehzangi (✉) · S. Alidoust
School of Architecture, Design and Planning, The University of Queensland, Brisbane, QLD, Australia
e-mail: a.chesh@uq.edu.au

A. Cheshmehzangi
School of Architecture and Urban Planning, Hunan University, Changsha, China

College of Architecture and Urban Planning (CAUP), Tongji University, Shanghai, China

W. Y. Chen
Department of Geography, The University of Hong Kong, Pok Fu Lam, Hong Kong

R. Fuller
School of the Environment, The University of Queensland, Brisbane, Australia

Centre for Biodiversity and Conservation Science, The University of Queensland, Brisbane, QLD, Australia

A. Cheshmehzangi et al. (eds.), *Nature in Cities, Nurturing Cities*, Urban Sustainability,
https://doi.org/10.1007/978-981-95-5487-4_1

Keywords Urban sustainability · Ecological integration · Resilient cities · Public health · Inclusivity · Placemaking · Climate adaptation · Nature · Nature in cities · Nature-based solutions · Nurturing cities

1 Framing the Challenge: Beyond Nature in Cities and Toward Innovative Integrated Design and Planning Solutions

Cities are at once humanity's greatest invention and its most pressing dilemma (Beauregard 2024). Though they also represent the cracks of unsustainable growth, e.g., air thickened with pollution, rivers forced into culverts, wildlife pushed to the outskirts, they also concentrate creativity, culture, and opportunity. The prevailing paradigm has historically positioned "the city" and "nature" on opposing ends of a spectrum, with the former being characterized by open space, wilderness, and ecological rhythms, and the latter by urbanization, infrastructure, and technology (Gandy 2022). This dichotomy has damaged the resilience of cities themselves in addition to degrading the way we plan and create our urban settings.

Recognizing that "nature in cities" cannot be viewed as ornamental, optional, or secondary is necessary to frame the problem. Urban trees are essential air quality and cooling infrastructure, not just a nice-to-have. As carbon sinks and flood barriers, wetlands are not residual land. Community gardens and parks serve as hubs for cultural memory, public health, and social inclusion in addition to being places for recreation. But all too frequently, these roles are disregarded when making planning decisions that are influenced by cyclical political cycles, short-term economics, or limited notions of efficiency.

The shortcomings of this fragmented strategy have been made clear by the mounting demands of our time. With rising sea levels and more intense heat waves transforming ecological degradation into existential risks, climate change is pushing the boundaries of current infrastructure. Public health emergencies show that ignoring nature is not just an environmental problem but also a human one, ranging from respiratory ailments associated with air pollution to mental health conditions made worse by a lack of green space. At the same time, cities replicate global patterns of inequality: access to safe, clean, and biodiverse environments is not uniformly distributed, resulting in a "green divide" (Tallent 2025) (Fobi Kontor et al. 2025) where the most vulnerable are frequently the ones who are most denied the benefits of nature.

A mindset change is necessary to move past this framing: cities must actively support natural areas as part of their fundamental identity and operational frameworks, rather than merely housing them. This entails integrating ecological concepts into the logic of urban planning itself, since the layout of streets, buildings, and public areas is inextricably linked to the movements of air, water, soil, and species. By combining urban design, landscape architecture, ecology, engineering, health sciences, and social policy into a more comprehensive practice, integrated design and planning solutions provide a means of achieving this goal.

Therefore, the question that must be answered is not whether cities can afford to incorporate nature, but rather if they can afford not to. Because floods, heat-related deaths, and other ecosystem service losses already cost billions of dollars a year, investing in ecological infrastructure is a practical way to address systemic urban risks rather than a luxury. Creative integrated approaches show that there are solutions when disciplines, scales, and communities come together. Examples include co-designed community gardens that combine social justice and ecological stewardship, green–blue corridors in flood-prone areas, and transit networks lined with trees that reduce emissions while increasing shade.

To frame the challenge of "nature in cities" in this book is to argue for a recalibration of urban futures. The dichotomy between built and natural must give way to a continuum where cities are conceived as socio-ecological systems. This calls for a shift in culture toward care—care for communities, care for ecosystems, and care for future generations, as well as political will and new forms of governance. The title of this book emphasizes the idea of nurturing, which emphasizes that integration is a continuous commitment rather than a one-time design effort. Similar to gardens, cities require care, perseverance, and a willingness to adapt.

There is a chance to drastically alter urban life by going beyond crude ideas of greening and instead promoting integrated design and planning solutions. Cities may develop into living innovation labs that are able to weather crises, promote health, and create a sense of belonging rather than being isolated technological displays. This book begins with that issue, focusing on nature's crucial role in creating resilient, inclusive, and compassionate urban futures rather than just its existence in cities.

1.1 Nature and the Urban Condition

The story of cities is often told as the story of humanity's separation from nature (Benton-Short and Short 2013; Turner et al. 2004). Historically, the term "urban" has connoted land dominance, ecological exploitation, and the creation of developed settings that impose order on wilderness (McMillan 2015). Ancient towns and cities were praised for their ability to create walls to keep the wild world at bay, clear forests for cultivation, and tame rivers with aqueducts. This reasoning was taken to its logical conclusion in nineteenth- and twentieth-century industrial cities, where rivers served as waste disposal channels and smokestacks rose as emblems of advancement. The dominant narrative never portrayed nature as an equal partner in forming urban life, but rather as a backdrop, resource, or threat.

The effects of this division are still evident today. Cities become fragile when urban expansion is planned against ecological systems. Floods move faster when rivers are channeled and straightened. Heat islands are exacerbated by paved landscapes that suffocate flora and soil. In addition to depriving urban residents of the precise ecological services—cooling, water purification, and food security—that are most critical during emergencies, the marginalization of wetlands and forests also diminishes biodiversity. In summary, the repression of nature has resulted in less resilient, egalitarian, and liveable urban ecosystems.

Yet, the story is not simply one of neglect. Creative interactions between natural processes and human design have also long occurred in cities. There are several examples of urban forms that worked with ecological rhythms rather than against them, such as the hanging gardens of Babylon or the rice terraces incorporated into Southeast Asian communities. Nature has always had supporters in urbanism, as seen by contemporary initiatives like Jane Jacobs's insistence on human-scaled streets enhanced by parks and trees (Al-Kodmany 2024) or Ebenezer Howard's Garden City concept (Richert and Lapping 1998). The issue is that these ideas were frequently marginalized and viewed more as idealistic concepts than as comprehensive plans for building cities.

The stakes are too great for nature to stay on the sidelines now. The occurrence of high temperatures, droughts, wildfires, and floods is a lived reality, not just a theoretical possibility. Public health emergencies, such as the correlation between pollution and asthma rates and the negative effects of concrete isolation on mental health, highlight the close connection between our surroundings and our overall health (Cheshmehzangi et al. 2025). The problem is made worse by inequality, since affluent areas may have lush parks and boulevards lined with trees, while underprivileged areas are frequently relegated to "grey deserts" (Gray 2008; Aldridge 2001), where there is little natural beauty and the greatest environmental dangers.

Therefore, to rethink the urban condition is to acknowledge that nature is a part of the city rather than something that exists outside of it. Hydrological cycles, biodiversity, air quality, and microclimate are all necessities for urban survival (Murugadoss et al. 2024); they are not frills. They undermine the basic pillars on which cities are constructed if they are disregarded. They serve as the framework for more resilient, inclusive, and healthy urban futures if they are accepted.

This is why nurturing cities begins with nurturing nature within them. The active verb "nurture" implies long-term stewardship, care, interaction and cultivation; it is more than just "preserve". Instead of just adding green spaces, a caring city fosters an environment that allows ecosystems to coexist peacefully with human populations. It guarantees equal access to nature's healing, uplifting, and protective properties for all locals, irrespective of their socioeconomic status or postal code (Sperry 2023). It celebrates diversity, including the variety of species and landscapes that enhance urban life in addition to the diversity of people.

Additionally, the urge to nurture reinterprets the function of design and planning. Cities must be viewed as living systems that require care and adaptation rather than as machines in need of technical repairs. This viewpoint necessitates integration at several scales: a restored wetland at the city's edge functions as a migratory stopover site for birds, while a street tree that shaded a bus stop contributes to regional and global climate regulation. Cities can transition from reactive crisis management to proactive resilience building by integrating ecological thinking into the very fabric of planning decisions.

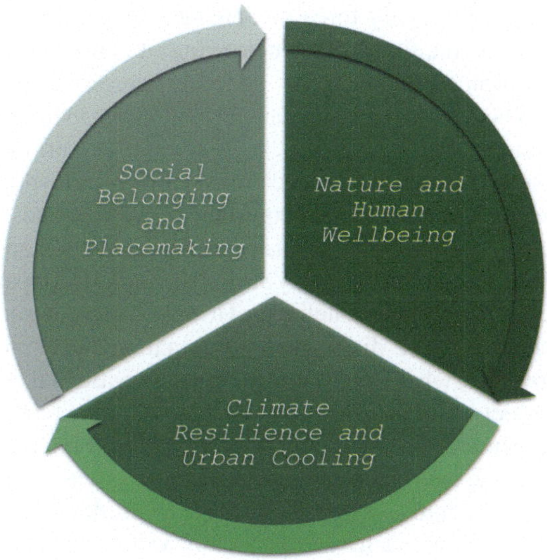

Fig. 1 Summary of three core parts of this book

Accordingly, the historical separation of city and nature is the barrier to sustainable urban futures and is not just a theoretical issue. To eliminate this divide is to envision an entirely different urban environment: one in which rooftops are transformed into meadows, rivers are not concealed in culverts but are hailed as urban lifelines, biodiversity thrives alongside cultural life, and the depth of ecological care, rather than the height of towers, is used to gauge progress.

This book takes that reimagining as its guiding principle. In a way, reinvention serves as the foundation for this work. It makes the case for a new paradigm of urban life by showing how nature can support resilience, placemaking, and well-being: cities that flourish because they are ecological and that nurture us because we nurture them. Future urban conditions must be ones of care, reciprocity, and integration, where nature is a resident rather than a tourist, and where nurturing emerges as the defining ethic of city-making. Hence, the book is divided into three parts of: (1) *Nature and Human Well-being*, (2) *Climate Resilience and Urban Cooling*, and (3) *Social Belonging and Placemaking* (See Fig. 1). We also note this is the first of three volumes we have put together under the topic of "nature in cities".

2 The Transformative Power of Nature

If the twentieth century was characterized by cities growing outward without considering ecological boundaries, the twenty first century is becoming increasingly marked by a realization that the ability of cities to successfully reintegrate environment into their fabric is essential to their existence, prosperity, and sense of justice. Nature is a revolutionary force that changes how cities operate, how people see them, and how communities define belonging; it is not merely a gentle ornament of the urban environment.

The most immediate transformation lies in health. It is indisputable from scientific evidence that having access to green space promotes mental health, lowers stress, lowers cardiovascular disease, and improves respiratory health. Youngsters who play beneath tree canopies grow more resilient, and senior citizens who live close to parks exhibit greater levels of social interaction and physical exercise. Urban nature became a lifeline for many during the COVID-19 pandemic, providing both safety and resilience against loneliness. However, there is still a glaring disparity in how these advantages are distributed. Marginalized areas are frequently relegated to "green poverty", where health inequalities increase and concrete predominates. Cities must put equity first to fully utilize nature's transformative power, making sure that all citizens have regular, safe, and meaningful interactions with biodiverse landscapes.

In the face of climate stress, nature is just as transformative. The limitations of gray infrastructure are exposed by heat waves, flooding, and storm surges, whereas green–blue systems, such as urban forests, wetlands, mangroves, and permeable landscapes, provide flexible and adaptable solutions. By lowering local temperatures by a few degrees, a tree canopy can prevent heat-related deaths. Compared to artificial drains, restored wetlands are significantly more effective in absorbing runoff, minimizing flood damage and storing carbon. Vulnerable communities are protected by coastal ecosystems, which serve as natural barriers against storm surges at a fraction of the expense of physical barriers. This is quantifiable resilience that saves resources and lives, not sentimental idealism. In this way, ecological systems redefine what is necessary for urban safety, functioning as active infrastructure rather than passive amenities.

Moreover, nature transforms inclusivity by providing shared areas where distinctions dissolve, going beyond resilience and health (Welden et al. 2021). Community forests, parks, and gardens transform into democratic spaces where social hierarchies are loosened and shared identities are created. By fostering a feeling of community via gardening, shared care of urban gardens in multicultural cities can help immigrant communities feel more connected to local customs. Expanding who belongs in the city and whose voices influence its future is made possible by framing access to nature as a right rather than a privilege.

Another aspect of change is provided by placemaking. In addition to being useful, nature is also cultural, symbolic, and closely related to how people perceive their surroundings. Identity is shaped by the stories and meanings that are carried by trees along a historic boulevard, a river slicing through a metropolis, or native vegetation recovering industrial land. We are reminded that cities are dynamic ecologies where human and non-human life coexist through biophilic design, ecological art, and even the existence of urban wildlife. Cities may foster attachment, pride, and continuity by incorporating nature into placemaking.

Therefore, the force of nature transcends all fields and disciplines. It necessitates a multidisciplinary strategy that connects government, ecology, public health, urban planning, climate science, and cultural studies. The case examples in this book, which range from coastal restoration in flood-prone areas to urban forests in dense Asian megacities, from high-tech ecological monitoring systems to community-driven gardens in low-income neighborhoods, show that innovation occurs when disciplines come together. These anecdotes demonstrate that integration is not merely theoretical; it is currently taking place and transforming urban futures in various settings.

This book embraces that ethos. In the urban drama, nature is shown as the protagonist rather than the backdrop. It encourages readers to view nature as the cornerstone of inclusivity, the scaffolding of resilience, the infrastructure of health, and the very fabric of location. The transformational force of nature is unleashed when cities are nurtured; this is not an afterthought, but rather the concept that sustains and reimagines urban life.

3 Purpose and Design of This Collection

This collection aims to reinvent the urban future through ecological integration and a caring philosophy, which is both intellectually and practically driven. All too frequently, urban discourse on nature is fractured, split up among disciplines, or boiled down to discrete interventions. That fragmentation is rejected in this book. Instead, it contends that fostering cities must start with fostering the natural systems that make them habitable and places environment as the fundamental infrastructure of urban life, influencing resilience, inclusion, identity, and health. By providing both critical analysis and concrete methods for incorporating nature into the routine operations of urban planning and architecture, the conceptual goal is to close the gaps between theory and practice.

The pragmatic objective is as aspirational. Social separation, health disparities, and climate change are challenges that cities around the world are facing. These interconnected problems are not being adequately addressed by traditional solutions, which are dominated by technology or simply commercial logics. This volume offers practitioners, academics, and decision-makers an evidence-based, innovative, and policy-relevant collection of ideas by curated viewpoints from many fields and situations. It is a forum where creative case studies, grounded techniques, and thoughtful tales show how cities can be fostered through ecological thinking rather than a manifesto of abstract ideas.

This collection also marks the beginning of a larger intellectual journey. *Nature in Cities, Nurturing Cities* is the first of three volumes dedicated to exploring the relationship between nature and cities—but more importantly, focused on nature "in" cities. Together, the trilogy aims to provide a thorough corpus of research and application that reinterprets urbanization via ecological integration. The foundation is laid in the first book, which emphasizes the crucial connections between human well-being, climate resilience, and placemaking. This narrative reflects well on our three core parts of the book, in which, several case studies are included. The discussion will be expanded to the other two volumes, where we will take a forward-looking approach, building future scenarios and speculative design to envision how cities can change when nature becomes an integral part of their character rather than a peripheral one. The trilogy is intended to be read as a continuity rather than as three separate books, with each volume building on the one before it to produce a comprehensive vision of nature-driven, nature-led, nature-first, and nature-based urbanism.

The collection's framework for this first book demonstrates a conscious dedication to both breadth and depth, bringing together a variety of viewpoints under three main thematic pillars. The first, "Nature and Human Wellbeing", examines the complex ways that social cohesiveness, mental equilibrium, and physical vigor are impacted by interactions with biodiversity, green spaces, and natural landscapes. It explores the evidence that nature is a vital factor in determining human health and quality of life, influencing behavior, creating connections, and bolstering psychological resilience, rather than just serving as a backdrop to urban life.

Examining how ecosystems and ecological infrastructures function as potent barriers against climate extremes is the focus of the second part of the book, i.e., "Climate Resilience and Urban Cooling". Nature works as a frontline ally to shield cities from growing environmental threats by reducing heat, absorbing floodwaters, improving air quality, and stabilizing local microclimates. In addition to providing defense, these systems pave the way for adaptive design by directing planners toward approaches that are predictive, regenerative, and resilient to unforeseen events. The third part, "Social Belonging and Placemaking", shows how urban settings may become hubs of shared memory, culture, and identity through the use of nature-inspired design. Here, ecological thought and creativity come together to demonstrate how natural features may support inclusive public places, anchor social life, and develop a feeling of common stewardship.

Every part aims to strike a balance between theoretical frameworks and grounded, real-world case studies, showing how integrating nature into cities is a daily practice, a policy framework, and a theoretical lens. These three threads collectively constitute the intellectual underpinnings of this volume, laying the groundwork for the two books that follow in the series: one that will explore the more general ideas and uses of nature-based urbanism, and another that will promote methods for designing nature into cities as a key component of ongoing and upcoming urbanization.

At its heart, this book is a reflection on responsibility. To nurture cities is to acknowledge that urban futures are inseparable from ecological futures. The design of this collection, and of the larger series it inaugurates, rests on the conviction that caring for cities requires caring for nature, not as an optional amenity, but as the very condition of survival and flourishing.

4 Toward a Narrative of Nurturing Cities

When we talk about nurturing cities, we are talking about a different ethic for urban life, one that centers our future vision and development around reciprocity, care, and interdependence. For far too long, cities have been thought of as growth or consumption machines, and their success has been gauged by their physical growth and economic output. The living systems that support cities, however, have been ignored, dispersed, or destroyed beneath the glass and concrete. Now is the moment to change this course. According to the narrative of nurturing cities, it is impossible to separate the ecological and urban conditions, and to plan for one without the other is to design for failure.

This book positions itself within a global discourse at a moment of profound urgency. Converging challenges such as climate change, biodiversity loss, and growing inequality call for a shift in values rather than small, gradual changes. In urban areas, nature is becoming an issue of survival, justice, and dignity rather than aesthetics or amenity. Developing cities means creating habitats where both human and non-human life may coexist peacefully, where resilience is built not just with steel and concrete but also with soils, marshlands, forests, wetlands, and the silent perseverance of species. It is to accept cities as dynamic ecosystems as opposed to static structures.

The vision presented here is both pragmatic and intellectual. It takes vision, patience, and humility to nurture. It acknowledges that cities need to be tended like gardens, with cycles of development, deterioration, and rebirth. It recognizes that advancement is not solely determined by monetary prosperity but also by the well-being of ecosystems, the inclusiveness of communities, and the ability to adapt responsibly. This book aims to reimagine what it means to build inclusive, resilient, and sustainable cities by redefining nature as necessary infrastructure and elevating care as a guiding principle.

This is only the beginning of a larger journey. As the first volume in a trilogy dedicated to *Nature in Cities* (and not just "Nature and Cities"), it lays the foundation for a deeper, continuing exploration. What follows in the chapters ahead are stories, observations, insights, and visions that make tangible the idea of nurturing cities, not as distant utopia, but as a practical, urgent, and achievable future. To nurture is to hope, and to hope is to act. The task before us is clear: to build cities that are not only liveable, but profoundly alive.

References

Aldridge ME (2001) WB Yeats and James Joyce: creating a unified Irish literary tradition. The University of Tulsa, Oklahoma

Al-Kodmany K (2024) Humanizing the high-rise city: podiums, plazas, parks, pedestrian networks, and public art. Routledge, New York

Beauregard RA (2024) Cities in the urban age: a dissent. University of Chicago Press, Chicago

Benton-Short L, Short JR (2013) Cities and nature. Routledge, New York

Cheshmehzangi A, Allam Z, Wang X, Su Z (2025) Integrating blue zone principles into healthy city design. In: Designing healthy cities: integrating climate-resilient urbanism for sustainable living. Springer Nature Singapore, Singapore, pp 75–88

Fobi Kontor M, Brown A, Núñez Collado JR (2025) Climate justice and heat inequity in poor urban communities: the lens of transitional justice, green climate gentrification, and adaptation praxis. Urban Sci 9(6):226

Gandy M (2022) Natura Urbana: ecological constellations in urban space. MIT Press, Cambridge, MA

Gray MA (2008) The traditional wilderness conception, postmodern cultural constructionism and the importance of physical environments. https://scholarworks.umt.edu/etd/1095/

McMillan BT (2015) "The tonic of wildness": Thoreau's critique of industrial capitalism. East Carolina University, Greenville, NC

Murugadoss D, Singh H, Thakur P (2024) Urban forests and microclimate regulation. In Urban forests, climate change and environmental pollution: physio-biochemical and molecular perspectives to enhance urban resilience.: Springer Nature Switzerland, Cham, pp 531–550

Richert ED, Lapping MB (1998) Ebenezer Howard and the garden city. J Am Plann Assoc 64(2):125–127

Sperry AB (2023) Nature and healing in urban communities: bioethical analysis of health and green spaces. http://hdl.handle.net/20.500.12613/8492

Tallent T (2025) A green divide? Climate policy support and its rural geography in Europe. West Eur Polit 1–31

Turner WR, Nakamura T, Dinetti M (2004) Global urbanization and the separation of humans from nature. Bioscience 54(6):585–590

Welden EA, Chausson A, Melanidis MS (2021) Leveraging Nature-based Solutions for transformation: reconnecting people and nature. People and Nature 3(5):966–977

Nature and Human Wellbeing

Nature and Human Health: Exploring the Vital Connection for Wellbeing

Reza Farhadi, Hadi Soltanifard, Fatemeh Mirzaeipour Meybodi, Maryam Noroozi, and Saeedeh Nasehi

Abstract The relationship between nature and human health has garnered considerable scholarly attention in recent years as urbanization and contemporary lifestyles increasingly distance individuals from natural environments. This chapter examines the multifaceted ways in which nature influences physical, mental, and emotional wellbeing, emphasizing the importance of incorporating natural elements into daily life to promote holistic health. Research consistently substantiates that exposure to nature whether through green spaces, forests, urban parks, or aquatic environments mitigates stress, elevates mood, and enhances cognitive function. These benefits are attributable to mechanisms such as decreased cortisol levels, heightened physical activity, and the restorative properties of natural landscapes in alleviating attention deficits and mental fatigue. The chapter begins by examining the physiological effects of nature, including its ability to lower blood pressure, enhance immune function, and promote physical activity. It subsequently discusses the psychological advantages, such as the diminution of symptoms associated with anxiety, depression, and stress-related disorders. Case studies and empirical evidence demonstrate the efficacy of nature-based interventions, such as forest bathing and green prescriptions, in optimizing mental health outcomes. Moreover, the social dimensions of nature are explored, focusing how communal green spaces facilitate social cohesion, alleviate loneliness, and enhance community wellbeing. Nevertheless, barriers to accessing

R. Farhadi (✉)
Department of Human Geography and Planning, Faculty of Geography, University of Tehran, Tehran, Iran
e-mail: reza.farhadi@ut.ac.ir

H. Soltanifard
Department of Environment, Hakim Sabzevari University, Sabzevar, Iran

F. M. Meybodi
Department of Landscape Architecture, Shahid Beheshti University, Tehran, Iran

M. Noroozi
Department of Environmental Design Engineering, University of Tehran, Tehran, Iran

S. Nasehi
Department of Environmental Planning, Management, and Education, Faculty of Environment, University of Tehran, Tehran, Iran

A. Cheshmehzangi et al. (eds.), *Nature in Cities, Nurturing Cities*, Urban Sustainability,
https://doi.org/10.1007/978-981-95-5487-4_2

nature, including urbanization, socioeconomic inequities, and environmental degradation, remain prevalent. This chapter is divided into three main sections. The first section explores the relationship between humans and nature, focusing on the physical, physiological, and cognitive dimensions. It provides examples to illustrate each of these dimensions. The second section presents policies and practical applications related to health and its various dimensions within the context of urban landscape planning and design. Finally, the third section identifies research gaps in this field and outlines a roadmap for future research. The chapter also advocates for equitable access to green spaces and the integration of nature into urban planning and public health frameworks. It also highlights the necessity for further research to clarify the long-term impacts of nature on health and to develop scalable, nature-based interventions for various populations. In conclusion, this chapter advocates for a transformative shift in recognizing nature as a vital pillar of public health. By prioritizing the incorporation of natural environments into urban design, healthcare systems, and community programs, societies can leverage the therapeutic potential of nature to foster resilience, wellbeing, and a sustainable future.

Keywords Nature and health · Green spaces · Mental wellbeing · Public health · Urban forest

1 Introduction

Today, over half of the global population resides in urban areas, with projections indicating that the urban population will nearly double by 2050, accounting for approximately seven in ten individuals (World Bank, 2023). However, the accelerated pace and extent of urbanization pose significant challenges (World Bank, 2023). This pronounced trend toward urban living has fundamentally reshaped human-nature interactions, resulting in a profound shift in our relationship with the natural environment. In a relatively brief period, humans have transitioned from predominantly outdoor lifestyles to one primarily conducted indoors. Currently, approximately 90% of the time is spent indoors in regulated environments, with an increasing trend of digital 'screen time'. For instance, over 80% of the UK population resides in urban areas, many of whom are estranged from nature. Global urbanization is diminishing access to nature in cities, which harms biodiversity, opportunities for nature engagement, and public health and wellbeing (Hedblom et al. 2024). The diminution of nature experiences may undermine individuals' emotions, attitudes, and behaviors toward nature, perpetuating a vicious cycle that exacerbates the consequences of this phenomenon. This disconnection can precipitate conflicts among stakeholders with divergent values (Ives et al. 2014). The growing alienation from nature is deeply troubling, as human-nature interactions significantly influence wellbeing, affinity for nature, and biodiversity conservation. Moreover, societal detachment from nature is a critical driver of unsustainability. Unrelenting environmental degradation could profoundly impact the quality of life, not only materially but also psychologically and spiritually. Humans derive substantial benefits from nature's contributions, both tangible and intangible. Numerous scholars argue that strengthening the human-nature bond is crucial, as increasing detachment from the environment exacerbates the global ecological crisis (Riechers et al. 2022). In urban regeneration, expanding urban greenery and enhancing access to green spaces while concurrently increasing urban density is not incompatible but a feasible strategy, as demonstrated by regeneration projects in cities such as Barcelona and Singapore. Furthermore, reconnecting individuals with nature can play a pivotal role in addressing contemporary ecological and sustainability challenges. The human-nature relationship has been scrutinized since antiquity; however, a heightened awareness of environmental crises in recent decades has spurred research (Ives et al., 2018). Numerous studies have explored the human-nature connection—the positive bond between individuals and the natural environment (Barrable and Booth 2022).

Increasing attention in fields such as environmental psychology, education (Chawla, 2020), and mental and physical health, and sustainability science(Ives et al. 2018; Wicks et al. 2022; Barragan-Jason et al. 2023). In this chapter, we adopt the term "human-nature connection" (HNC) as an overarching concept, encompassing diverse terms across disciplines. A healthy city is attentive to the health of its residents and strives to enhance it (Hedblom et al. 2024).

1.1 Scope and Significance of Integrating Nature into Health Frameworks

There is growing recognition of the necessity for daily engagement with green spaces and natural environments to lead a fulfilling, productive, and meaningful life (Lehmann 2023). Various terms related to nature connections have emerged from diverse disciplinary perspectives and normative agendas. For instance, the "biophilia hypothesis" posits that humans possess an innate affinity for nature, a paradigm that underpins much scholarly and applied work to foster interactions with green environments. Another concept, "nature deficit disorder", describes the adverse developmental impacts of diminished outdoor engagement on children. Similarly, the "extinction of experience" refers to the phenomenon whereby urbanization curtails everyday experiences of nature, with implications for health, emotions, attitudes, and behavior (Ives et al. 2018). Human-nature connectedness is a multidimensional concept, with recent studies identifying five distinct dimensions: (1) the material dimension (e.g., food, fuel, or handcrafted goods); (2) the experiential dimension, encompassing nature visits and outdoor activities; (3) the emotional dimension, including spirituality, aesthetics, and sense of place; (4) the cognitive dimension, related to knowledge and awareness of nature; and (5) the philosophical dimension, concerning humanity's relationship with the natural world. Additional factors, such as sociocultural, institutional, and political influences, modulate these dimensions. Figure 1 depicts the dimensions of disconnection from nature. These dimensions operate along a continuum, from external connections (e.g., physical interaction) to internal connections (e.g., emotions or worldviews). Moreover, analyzing the scales (individual and societal) at which these connections function is essential (Ives et al. 2018).

The study of nature exposure and health outcomes has significantly expanded in recent years. Restoring the human-nature connection could serve as a critical pathway to enhance both the physical and psychological wellbeing of individuals and communities while concurrently fostering personal and collective environmental stewardship (Zelenski et al. 2023). Familiarity with nature and species identification skills are notably higher among individuals who currently reside or have resided in rural areas during childhood, exhibiting greater nature-relatedness compared to their urban counterparts. Consequently, diminished opportunities for human-nature interaction weaken cognitive and emotional connections to the natural world. Such reductions can adversely affect preferences for human-nature relationships, perpetuating a pervasive negative cycle that undermines relational values—such as care for nature, sense of belonging, place attachment, and identity—that underpin both human wellbeing and environmental stewardship (Bashan et al. 2021). A study investigated how employing cues to experience nature as a mechanism

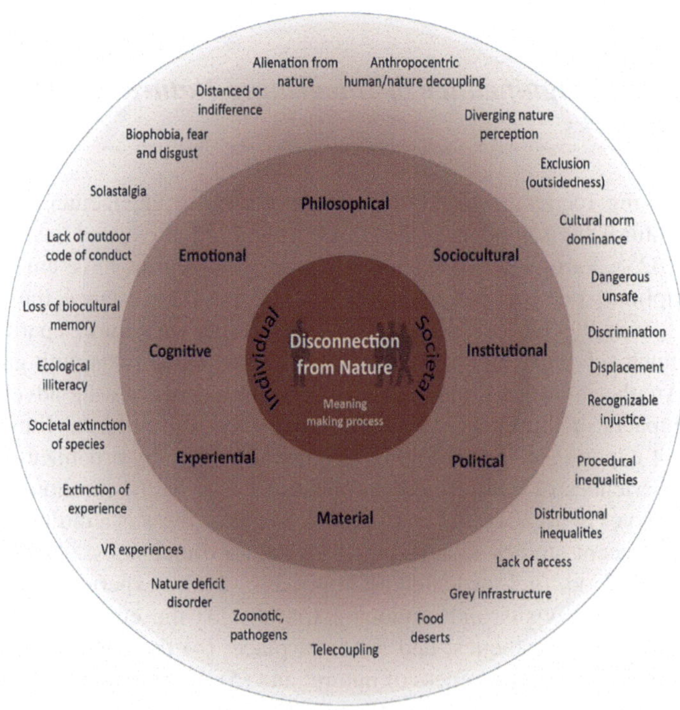

Fig. 1 Dimensions of disconnection from nature, Adopted from: (Beery et al. 2023)

for inducing situational nature relatedness (NR) can enhance the quality of individuals' interactions with green spaces and their post-visit positive affect. Participants who received cues promoting close psychological proximity to nature (e.g., smelling and touching natural elements) interacted with nature three to four times more frequently and reported 0.2 units higher positive affect compared to other participants, suggesting that cues fostering closer engagement with nature can significantly enhance the quality of natural interactions and elevate positive affect (Colléony et al. 2020). Broadly, studies have conceptualized reconnecting individuals with nature as a therapeutic intervention, often targeting the individual level. Reconnecting with nature extends beyond mere physical interaction, encompassing the deliberate cultivation of cognitive, emotional, and biophysical connections that positively shape human-nature interactions(Ives et al. 2018). Plant and animal biodiversity contribute significantly to human health across multiple dimensions.

2 Physiological Benefits of Nature Exposure

2.1 Impact on Cardiovascular Health

Compelling evidence suggests that engagement with natural environments enhances heart rate variability, improves mental health and executive function, and reduces the risk of cardiovascular disease and obesity (Wen et al. 2019; LoTemplio et al. 2023). Green spaces may mitigate overweight and obesity primarily by promoting physical activity; however, research findings on this relationship are inconsistent. Some studies indicate that exposure to green spaces is associated with lower obesity rates in children and adults, while others report no significant association or, in some cases, an elevated risk of being overweight in areas with greater greenery (Jimenez et al. 2021). Lower adiposity levels are associated with greater availability and proximity to green and blue spaces, as indicated by 68% of studies (Teixeira et al. 2021). Nevertheless, few studies have explored external factors that may underpin this relationship. Robust evidence demonstrates that exposure to natural environments lowers blood pressure (Ye et al. 2023). A study examining the association between natural and built environments and children's blood pressure found that green spaces near residences and schools were correlated with lower blood pressure, potentially mediated by outdoor play, and a reduced risk of obesity (Dzhambov et al. 2022). Overall, research consistently indicates a direct link between physical activity and cardiovascular health. The sequence of walks also influenced cortisol levels, underscoring the stress-mitigating benefits of natural settings (Aras et al. 2024). However, lifestyle choices play a significant role in the development of cardiovascular diseases. Individuals who are obese face increased risks of chronic diseases, negative mental health outcomes, and early mortality, much of which can be attributed to their lifestyle and its effects. Additionally, external factors are crucial, with studies highlighting two main areas of concern: environmental health factors, such as air pollution, and social factors, including socioeconomic status and access to nutritious food (Lu et al. 2024).

2.2 Immune System Modulation and Inflammation Reduction

Exposure to phytoncides—volatile organic compounds emitted by plants and trees to protect against harmful insects and microbes—effectively enhances immune function by reducing stress hormones and increasing natural killer (NK) cell activity while elevating levels of anticancer proteins(Zingoni et al. 2017). Research on the association between green spaces and cancer is limited and varies by cancer type. For instance, an inverse relationship was observed between neighborhood greenery and lethal prostate cancer among sedentary individuals in high-density urban areas (Iyer et al. 2020). Another study demonstrated that sensory engagement with nature and participation in nature-based activities fosters psychological distance from negative emotions, thereby ameliorating the psychological burden of cancer diagnoses through the restorative effects of nature (Quah et al. 2024).

2.3 Promotion of Physical Activity through Green Spaces

The human-nature relationship in urban environments can be conceptualized through multiple dimensions: physical, visual, social, and psychoemotional. While physical interactions with nature in urban settings often follow predictable patterns, relationships at other levels are shaped by cultural, social, and economic factors. Physical activity is frequently the most prominent outcome of human-nature interactions, manifesting in daily life and serving as a foundation for additional benefits, such as enhanced social cohesion. Increased physical activity is a notable consequence of access to urban green spaces, encompassing diverse activities such as individual or group walks, cycling, social gatherings, and leisurely outdoor time. Empirical evidence substantiates a positive association between nature exposure and elevated physical activity levels (Jimenez et al. 2021). Physical activity functions as a critical mechanistic pathway for achieving beneficial health outcomes by facilitating opportunities for outdoor exercise (e.g., walking and play) (Warburton and Bredin 2017). A study comparing walking scenarios in distinct settings a 50-minute walk on a forest path, a 50-minute walk along a busy road, and routine daily activities found that walking enhances mental wellbeing compared to daily activities, regardless of the environment. However, forest walks yielded the most substantial and sustained improvements in mental state (Koselka et al. 2019). Robust evidence confirms that regular physical activity reduces the risk of premature mortality and serves as an effective preventive measure for at least 25 chronic health conditions (Warburton

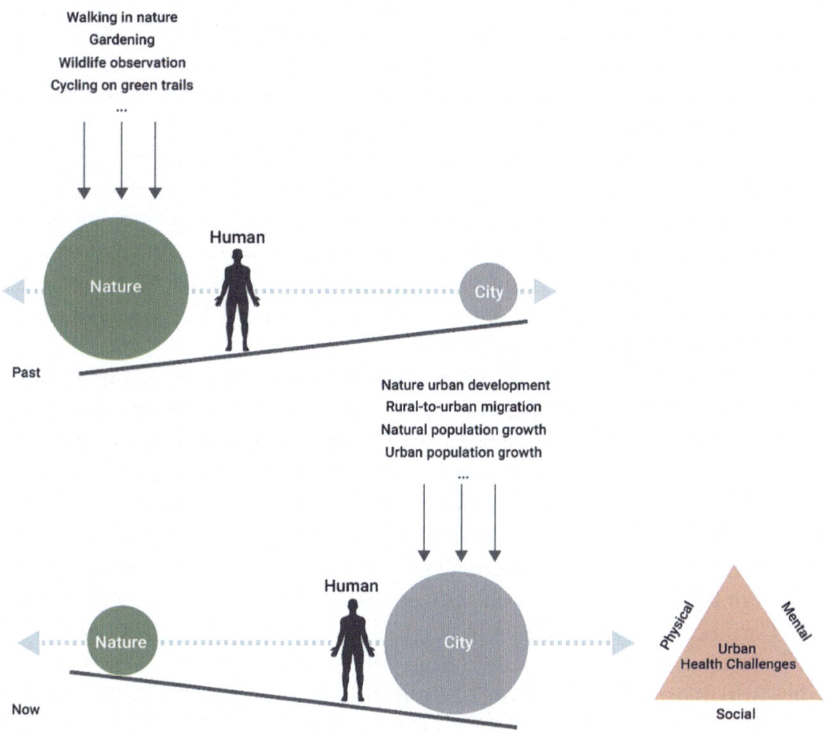

Fig. 2 Transformation of the Interconnection between People, Natural Environments, and Urban Spaces

and Bredin 2017). Figure 2 illustrates the evolution of the relationship between urban environments and nature with focusing on urban health.

2.4 *Socioeconomic Inequities in Access to Green Spaces*

Access to nature is unevenly distributed across different socioeconomic and demographic groups. Urban areas often exhibit pronounced socioeconomic inequalities that manifest as spatial and physical disparities within cities (Sarkar et al. 2024). These disparities are particularly evident in the differential access various social groups have to urban services and infrastructure, with access to urban green spaces serving as a notable example. Numerous studies have demonstrated that social groups—including residents of deprived neighborhoods, racial and ethnic minorities, religious minorities, and women—experience varying degrees of access to green infrastructure due to these inequalities. For instance, The Health Foundation (2024) conducted a study investigating disparities in access to green space in the United

Kingdom.[1] Their findings revealed that residents in more deprived areas are significantly more likely to live in neighborhoods with limited access to green space. Specifically, 28% of residents in the most deprived neighborhoods reside within the 10% of neighborhoods with the least access to green spaces, compared to only 7% of individuals in less deprived neighborhoods (The Health Foundation, 2024). Similarly, a study conducted in Germany by Wüstemann et al. (2017) reported that 92.8% of the German population has access to green space within a 500-meter radius of their homes. On average, urban residents in large German cities live approximately 183 meters from the nearest green space and have access to 4.4 hectares of urban green space within that distance, equating to an average of 8.1 square meters per person (Wüstemann et al. 2017).

Research consistently shows that low-income communities and communities of color have both less green space within their neighborhoods and fewer nearby parks. In a study conducted across five US cities, Williams et al. (2020) identified significant disparities in access to urban parks, particularly in terms of safety concerns. These inequalities were primarily linked to racial, ethnic, and economic factors, with low-income communities experiencing the lowest levels of park accessibility (Williams et al. 2020). The Health Foundation's 2024 study further highlighted ethnic disparities in the UK, reporting that 40% of individuals identifying as Black, Black British, Caribbean, or African live in neighborhoods with the least access to green space. In comparison, 37% of people from "other" ethnic groups, 34% from Asian or Asian British backgrounds, and 27% from mixed or multiple ethnic groups reside in similarly deprived areas. Only 13% of individuals from white backgrounds live in neighborhoods lacking green space. Conversely, just 5% of Black, Black British, Caribbean, or African individuals live in neighborhoods with the greatest access to green space, compared to 8% of those from Asian, Asian British, or "other" ethnic backgrounds, 15% from mixed or multiple ethnic groups, and 27% of white individuals (The Health Foundation, 2024). Additional barriers to equitable access include economic factors such as entry fees for public parks, which disproportionately affect low-income groups. Restrictions on gathering food and fodder further limit the access of some communities to traditional ecosystem services. Gender inequities also persist; many women report a lack of time and safety concerns as significant barriers to park visitation. Safety considerations can both encourage and deter park use. Focus group research has identified factors that promote park attendance, including the presence of authority figures, other visitors with similar intentions, secure parking, well-maintained trails, and adequate lighting. Conversely, violence, disruptive behavior, and a general sense of danger discourage park visits. These safety concerns disproportionately affect communities already facing limited access to natural spaces (Huang and Napawan 2021; Polko and Kimic 2022). Research focusing on Muslim communities' access to natural spaces in the United Kingdom has revealed multiple layers of exclusion. These include a lack of inclusive imagery in promotional materials, inadequate facilities for social gatherings, prior experiences of discrimination, perceptions that protected areas are predominantly white spaces,

[1] For more information, please see the website: https://www.health.org.uk/

and general unfamiliarity with these environments (Edwards et al. 2023). Furthermore, inequalities in urban communities have been exacerbated by local and global events, notably the COVID-19 pandemic. During the pandemic, certain communities experienced disproportionately high case rates, underscoring the intersection of environmental justice and public health. Areas predominantly inhabited by people of color not only exhibited elevated COVID-19 incidence rates but also had less greenery. Research found that a 0.1 increase in the vegetation index corresponded to a 4.1% decrease in COVID-19 incidence rates (Spotswood et al. 2021).

2.5 *Mechanisms: Stress Reduction (Cortisol Regulation) and Restorative Environments*

Kaplan (1998) asserts that regular engagement with nature enhances wellbeing. Proximity to natural environments correlates with improved physical and mental health, and participation in nature-based activities supports effective functioning (Kaplan 1998). Robust evidence suggests that nature engagement confers health benefits; however, uncertainty remains about whether these mechanisms operate independently or synergistically, as most studies examine them separately (Nieuwenhuijsen et al. 2014). Key attributes (e.g., biodiversity, disturbance level, accessibility) driving beneficial interactions and their variation across cultures and socioeconomic groups are unclear (Keniger et al. 2013). The association between natural outdoor environments (NOE) and health varies by factors such as contact type, life stage, and activity level. Nature exposure alters prefrontal cortex activity, critical for emotion regulation (Hansen et al. 2017). Despite strong evidence, mechanisms of urban green and blue spaces (UGBS) remain partially understood and differ across subgroups (Hunter et al., 2023). Table 1 represents the aspects of UGBSs in urban environment.

A study highlights three key mechanisms of NOE health benefits: physical activity, social interaction, and stress reduction, with mental wellbeing as a proxy. Use and experience (e.g., perceived greenness, satisfaction) consistently correlate with these outcomes, unlike NOE quantity or quality, which vary by context (Kruize et al. 2020). Understanding UGBS mechanisms—promoting physical activity, fostering social interaction, visual engagement, and reducing pollution or urban heat—is crucial. Physical activity is a pivotal mediator (Hunter et al., 2023). NOE health benefits stem from: (1) reducing mental stress, aiding recovery from attentional deficits; (2) promoting physical activity; (3) facilitating social contact; (4) supporting child development; and (5) enhancing purpose (Kruize et al. 2020). Preferences are shaped by attributes like naturalness, tranquility, safety, and accessibility (Shahamati 2020). Extending Markevych et al.'s (2017) pathways, blue spaces involve:

- Mitigation (e.g., reducing urban heat);
- Instoration (e.g., enhancing mood, activity);

Table 1 Pathways linking green and blue spaces to health

way	Aspect	Benefits and Effects
Mitigation (harm reduction)	Urban heat island	Urban blue spaces contribute to temperature regulation by absorbing heat during the day and releasing it at night, thereby reducing heat-related mortality among vulnerable populations residing near water bodies. Similarly, urban green spaces mitigate the urban heat island effect through multiple mechanisms, including shading provided by tree canopies, absorption of solar radiation by vegetation, and cooling via evapotranspiration. These processes collectively modify the urban thermal environment, leading to reductions in air temperature, alleviation of heat stress, and decreased heat-related mortality.
	Air pollution	Exposure to traffic-related pollution is reduced in areas characterized by the absence of primary emissions and the presence of mitigating features such as street trees, which decrease near-road pollution exposure. Vegetation contributes to the removal of air pollutants, including particulate matter (PM10) and ozone, through processes of deposition. Consequently, air quality is improved in the vicinity of residences and schools, promoting healthier living environments.
	Noise	Water bodies tend to increase environmental sound levels, whereas green spaces typically absorb sound and help mitigate noise pollution in urban areas. Although blue spaces may elevate ambient sound, they can enhance health outcomes by masking unpleasant noises such as traffic. Experimental studies have demonstrated that water sounds, either alone or combined with birdsong, reduce stress more rapidly than urban noise, silence, or even calming music.
	Aerosols and negative ions	Exposure to environments with high-density negative ions has been associated with improved lung function and reduced inflammation, with these benefits persisting for several months after exposure. Additionally, lower depression scores have been observed in individuals exposed to such conditions. Compared to other alpine settings, these environments are linked to reduced stress levels and decreased breathing difficulties, particularly among children with asthma.

(continued)

Table 1 (continued)

way	Aspect	Benefits and Effects
	Aerosolised toxins	Natural water bodies can aerosolize microorganisms, some of which produce toxins that may be harmful to respiratory health. However, marine aerosols at low concentrations have been suggested to confer positive health effects. For example, yessotoxin, a marine biotoxin, has demonstrated potential in reducing inflammation and enhancing immunoregulation.
	Solar irradiance	Increased exposure to ultraviolet (UV) radiation can elevate the risk of skin cancer. However, UV exposure also facilitates the synthesis of vitamin D, which has been associated with a reduced risk of autoimmune disorders, cardiovascular diseases, certain cancers, and poor mental health.
Instoration	Encouraging physical activity	Physical activity levels tend to be higher near coastal and waterside locations. Moreover, individuals engage in longer exercise durations in blue spaces compared to green or urban environments.
	Supporting positive social relations Facilitating social cohesion	Blue spaces have been shown to promote positive social relationships more effectively than green spaces. Access to blue spaces is associated with higher levels of social support, whereas urban green spaces primarily facilitate neighborly contact and increased social cohesion. Green spaces also enhance residents' sense of safety and mutual assistance. Both types of environments contribute to improved mental and general health, highlighting the significant influence of the physical environment on wellbeing.
	Building place/nature connectedness	Individuals residing along the coast tend to exhibit higher levels of urban nature connectedness. Humans possess a profound, innate need to feel connected to others or to something greater than themselves. This sense of connection acts as a protective buffer against poor mental health and helps mitigate the risks associated with loneliness. Moreover, it enhances emotional wellbeing in blue space environments and fosters a stronger, enduring trait of nature connectedness.

(continued)

Table 1 (continued)

way	Aspect	Benefits and Effects
Restoration	Restoration (capacity restoration)	Lowering stress levels is essential because individuals require opportunities to recover and "recharge" their emotional and cognitive resources. According to Stress Reduction Theory (Ulrich et al., 1991), such recovery is necessary to alleviate the demands placed on emotional resources. Similarly, Attention Restoration Theory (Kaplan & Kaplan, 1989) emphasizes the need to restore cognitive and attentional capacities through restorative experiences.
	Stress reduction	Numerous studies have consistently shown that blue spaces are among the most restorative environments. Exposure to aquatic scenes facilitates relaxation, thereby promoting faster recovery from stress. Increased levels of happiness have been observed in marine and coastal settings. Furthermore, walks in blue spaces are associated with enhanced wellbeing, yielding mood benefits comparable to those documented in green spaces. Additionally, subjective wellbeing is significantly improved through interaction with blue environments.
	Cognitive restoration	Better performance on a range of attentional tests following exposure to "natural" images and sounds, with improved attention and cognitive performance. Differentiating the cognitive effects of blue versus urban green spaces.

Source (Bell et al. 2021; Dzhambov et al. 2022; Farhadi et al. 2025)

3 Restoration (E.G., Recovering from Stress) (Bell et al. 2021). Psychological and Emotional Wellbeing

Active engagement with nature, coupled with emotional involvement and reflection on meaning, is correlated with nature connectedness, a psychological measure (Ives et al. 2018). Consequently, the psychological impact of disconnection in urban environments is substantial. Urban settings, characterized by dense infrastructure, noise pollution, and scarce green spaces, can exacerbate mental health issues. Urbanization is recognized to adversely affect mental health, primarily due to limited access to nature (Wang et al. 2024). During the COVID-19 pandemic, disparities became evident, as urban green spaces provided critical relief for mental health and safe recreation but were disproportionately inaccessible to certain communities (Spotswood et al. 2021).

3.1 Cognitive Restoration

The disparity between the natural environment, to which human physiological systems are adapted, and highly urbanized, artificial surroundings contributes significantly to the modern "stress state" (Song et al. 2016). Robust evidence demonstrates that natural environments confer substantial health benefits (Jimenez et al. 2021). One key mechanism is stress amelioration. A meta-analysis revealed that increased exposure to nature was associated with reductions in salivary cortisol, state anxiety, self-reported stress, systolic blood pressure (SBP), diastolic blood pressure (DBP), Ln(LF/HF) of heart rate variability (HRV), and enhanced restorative outcomes (Yao et al. 2021). Substantial evidence indicates that cortisol levels decrease in natural environments, with studies showing reduced salivary cortisol after mild-to-moderate exercise in natural settings compared to urban ones (Song et al. 2016). Exposure to nature mitigates the negative effects of stress. A study on wetlands demonstrated significant improvements in mental health, including psychological wellbeing, anxiety (as measured by the Generalized Anxiety Disorder-7), stress (as measured by the Perceived Stress Scale), and emotional wellbeing, by offering a sense of escape and facilitating relaxation. Nature exposure is linked to reduced anxiety and rumination, factors associated with depression and anxiety disorders. A systematic review and meta-analysis confirmed a reduction in depressive mood after brief exposure to natural environments (Keeler et al. 2019).

3.2 Cognitive Restoration and Attention Recovery

The biophilia hypothesis, proposed by Edward O. Wilson, posits that humans have an innate affinity for nature, shaped by evolution. Attention Restoration Theory (ART), grounded in this concept, elucidates how exposure to nature benefits mental health. ART asserts that the mental fatigue of modern life stems from a depleted capacity for directed attention (Kaplan 1998). Natural environments, rich in restorative qualities, enable recovery of this capacity(Kaplan 1998). Experimental studies support the protective effects of nature on mental health and cognitive function (Jimenez et al. 2021). A survey of University of Helsinki students revealed that outdoor recreation in nature significantly enhances wellbeing during stressful periods, such as the COVID-19 pandemic. Compared to urban settings, exposure to nature has been shown to improve attention, executive function, and perceived restrictiveness (Jimenez et al., 2021; Stenfors et al., 2019).

3.3 Empirical Evidence from Nature-Based Interventions

Nature therapy is increasingly recognized as a promising approach to preventive medicine (Song et al. 2016). As a health-promotion strategy, it mitigates the modern "stress state" and "technostress" (Hansen et al. 2017). A study demonstrated that nature therapy in the Linpan environment significantly benefits older women by regulating blood pressure and pulse, alleviating cardiovascular disease, and improving sleep quality, with pronounced effects in those with hypertension(Ye et al. 2023). Forest bathing yields multiple benefits, including improved cardiovascular function, hemodynamic, neuroendocrine, metabolic, immune, inflammatory, antioxidant, and electrophysiological indices; enhanced emotional state, attitudes, and psychological recovery; and significant reductions in anxiety and depression. These interventions substantially enhance physical and mental health (Wen et al. 2019). Table 2 Shows the other benefits of urban nature.

4 Policy and Practical Implications

In response to the intensification of environmental challenges over recent decades, various strategies have been proposed to mitigate these issues. Among the most significant practical approaches is the integration of design and planning with nature, which has effectively addressed concerns related to climate change, enhanced adaptation and resilience, promoted public health, combated global warming, and supported biodiversity conservation. Nature-based solutions (NBS) have garnered considerable attention from urban planners, designers, and policymakers during the past decade. This approach, grounded in a flexible conceptual model, is based on emulating the forms, processes, and mechanisms of natural systems and incorporating them into the built environment (Zhong et al. 2024). Consequently, NBS has facilitated practical interventions that preserve ecological values while promoting their sustainable and balanced use.

One prominent practical approach to urban development is biophilic design. The term "biophilia", first introduced by Edward O. Wilson in 1984, denotes the innate human affinity for nature (Thomson and Newman 2021). Over time, this concept has evolved beyond a design strategy into a broader social movement and urban policy initiative aimed at embedding nature within cities and urban planning across multiple spatial scales. These solutions harness scientific, technical, and engineering expertise to replicate natural systems within construction and urban layouts(Newman et al. 2017).

In addition to biophilic design, green infrastructure planning represents another vital nature-based approach to addressing urban challenges. Green infrastructure involves the strategic planning and development of an interconnected network

Table 2 Benefits of urban nature

Benefit	Description	Examples
Physical health	Promotion of physical activity through green spaces	Exposure to outdoor environments offers a mechanistic pathway for improving health outcomes. Increased access to opportunities for outdoor exercise, particularly walking, is associated with enhanced mental wellbeing, especially within forest environments. Furthermore, regular physical activity demonstrably reduces the risk of premature mortality and plays a significant role in preventing at least 25 chronic health conditions.
	Impact on cardiovascular health (e.g., lowered blood pressure, heart rate variability)	Moreover, research suggests that exposure to green spaces is associated with a range of positive health outcomes. These benefits include a reduction in the incidence and severity of headaches, as well as support for addiction recovery (Keniger et al., 2013). Furthermore, natural environments encourage physical activity, thereby contributing to a reduced risk of cardiovascular disease and lower adiposity, which together help combat obesity. Residential and school green spaces have also been linked to lower blood pressure.
	Immune system modulation and inflammation reduction	Natural environments are believed to have a beneficial influence on both the immune response and psychological wellbeing. Specifically, studies have demonstrated that exposure to such environments is associated with enhanced immune function, a reduction in stress hormone levels, and increased activity of natural killer (NK) cells. Furthermore, elevated levels of anticancer proteins have been observed in individuals engaging with natural settings. Finally, the experience of natural environments may foster the development of psychological distance from negative emotional states.
Mental health	Nature's role in mitigating anxiety, depression, and stress-related disorders	Research indicates that exposure to natural environments is associated with a range of positive psychological and physiological effects. Studies have demonstrated a correlation between access to green spaces and increased self-esteem (Keniger et al., 2013; Béclin & Rusinek, 2024). Moreover, nature exposure has been shown to reduce stress biomarkers, including decreased salivary cortisol levels, lower state anxiety and self-reported stress, and a reduction in systolic and diastolic blood pressure (SBP, DBP). Additionally, improved heart rate variability, as measured by Ln(LF/HF), has been observed. Furthermore, immersion in natural environments is linked to lower levels of anxiety and rumination, decreased depressive mood after short-term exposure, and increased restorative outcomes.

(continued)

Table 2 (continued)

Benefit	Description	Examples
Cognitive benefits	Cognitive restoration and attention recovery (e.g., attention restoration theory)	Exposure to natural environments has been shown to promote cognitive restoration and enhance learning outcomes. Research suggests that such exposure can reduce mental fatigue and improve academic performance while also providing opportunities for education and learning. Furthermore, access to green spaces has been linked to improved task performance and enhanced cognitive function, particularly in children. Nature's restorative effects on attention capacity are well-documented (Keniger et al., 2013), and this restorative capacity offers protective effects on mental health and cognitive function, promoting emotional and cognitive renewal. These benefits are particularly salient during stressful periods, such as the COVID-19 pandemic. Additional benefits include improved attention, enhanced executive function, and increased perceived restorativeness.
Social benefits	Greater social cohesion and improved social relationships	Access to natural environments offers opportunities for both shared and solitary experiences, contributing to a range of positive social outcomes. Green spaces have been shown to facilitate social interaction and enable social empowerment. Research also suggests a correlation between access to natural areas and reduced crime rates and violence. Furthermore, these environments foster social cohesion (Shanahan et al., 2016) and provide opportunities for social support (Keniger et al., 2013; Shanahan et al., 2016).

of natural and semi-natural green and blue elements within urban areas, delivering a wide array of ecosystem services to mitigate diverse urban issues (Gómez-Baggethun and Barton 2013). While primarily a planning concern, green infrastructure also encompasses design solutions that promote sustainability and environmental resilience. A fundamental principle underpinning both planning and design efforts is the connectivity of infrastructure networks, which enhances ecosystem services through positive synergies among network components (Monteiro et al., 2020). Singapore exemplifies the successful application of biophilic design principles and green infrastructure planning in urban policymaking. One of the country's primary challenges has been managing open space within a high-density urban context. This challenge often appears paradoxical as planners' endeavor to reconcile resource-efficient city design with expansive ecological considerations. Nevertheless, the integration of green infrastructure networks with biophilic design principles has yielded numerous benefits, including increased biodiversity, improved air quality, reduced surface water runoff, enhanced energy efficiency, and urban cooling (Newman 2014).

Although urban planners traditionally prioritize housing, economic development, and social management, it is imperative to adopt an integrated perspective that recognizes the dynamic relationship between citizens and nature. Such an approach can serve as a cornerstone of practical urban planning measures. Ultimately, sustainable urban development can be achieved through the deliberate planning and development of green spaces and interconnected urban networks that foster ecological and social wellbeing. Implementing the aforementioned recommendations will enable policymakers and urban planners to advance the development of cities that are both more sustainable and more resilient(Farhadi et al. 2024).

5 Gaps, Future Directions, and Research Needs

The success of urban greening projects depends on comprehensive planning, active community engagement, and careful consideration of both the diverse benefits and potential challenges (Mattijssen et al. 2019). By integrating nature-based solutions and emphasizing environmental justice, cities can foster more sustainable, resilient, and equitable urban environments (Anguelovski et al. 2020). Nevertheless, research on human exposure to nature within urban settings continues to face significant challenges despite the well-documented health benefits. These challenges primarily relate to methodological approaches, measuring benefits, and reporting results, which complicate decision-making for urban planners and policymakers. Numerous studies demonstrate a direct association between increased exposure to urban green spaces and improvements in mental, physical, and certain social health outcomes among urban populations. This evidence is supported by both cross-sectional and an increasing number of longitudinal studies. However, the current evidence base is limited by inconsistencies in measurement, a scarcity of rigorous long-term or intervention studies, and substantial gaps concerning causal inference, equity, contextual diversity, and qualitative aspects of green space(Hartig et al. 2014; Geneshka et al. 2021; Geary et al. 2023). While many studies benefit from the use of up-to-date satellite data, advanced analytical tools such as Geographic Information Systems (GIS), and reliance on longitudinal datasets combined with conceptual frameworks, they also encounter significant limitations. These include the absence of standardized definitions for green spaces and health outcomes, resulting in methodological heterogeneity and challenges in synthesizing findings (Geneshka et al. 2021). Furthermore, there is an over-reliance on observational study designs and a lack of data concerning disadvantaged urban populations and ethnic minorities (Jimenez et al. 2021). Additional issues include variability in examining the health effects of blue spaces (water bodies), green space quality, biodiversity, and differences in the spatial and temporal scales of studies. Table 3 summarizes the identified research gaps, future directions, and critical research needs.

Table 3 The summary of the gaps/limitations and future direction and needs

Gap/Limitation	Future Direction/Research Need
Heterogeneity in exposure metrics and measurement methods	Standardize metrics for green and blue space exposure as well as health outcome assessment
Lack of longitudinal/intervention studies (overreliance on cross-sectional designs)	Conduct more longitudinal, prospective, and intervention studies (e.g., natural experiments) across diverse urban settings
Limited causal inference due to confounding/ self-selection	Apply rigorous analytical methods to strengthen causal inference, adjusting for socioeconomic status, self-selection, and mediators
Inconsistent or inadequate green/blue space quality and biodiversity assessment	Develop and implement standardized quality metrics, including tree canopy cover, usability, biodiversity, and safety
Limited study of blue space health impacts	Expand studies focusing on the distinct health effects of blue space exposure in urban environments
Sparse data on equity, socio-demographic moderators, and vulnerable subgroups	Prioritize research on diverse socioeconomic, ethnic, and disadvantaged groups; examine equity and accessibility effects
Lack of research in diverse global/urban contexts, including global south cities	Broaden geographic scope to include under-represented regions and a variety of urban forms, including cities in the global south
Insufficient integration of mechanistic/ biomarker/mediation analyses	Investigate and quantify biological and psychological mechanisms (e.g., stress recovery, immune function) and mediation pathways
Underutilization of advanced/exposure assessment technologies (wearables, high-res remote sensing)	Utilize and further develop novel exposure assessment tools and real-time tracking technologies such as LiDAR, GPS, and biomarker integration
Few economic evaluation and policy-relevant studies	Conduct studies assessing the economic value and cost-effectiveness of urban greening and blue space interventions
Minimal studies comparing types of greenspace interventions or with alternative city investments	Compare the effectiveness of green and blue space interventions with alternative urban improvement strategies
Absence of standardized reporting of negative or null effects	Encourage transparent reporting and analysis of null, mixed, or adverse effects, including issues like gentrification and poor-quality green spaces
Limited investigation of cumulative/ life-course effects and critical exposure periods	Examine dose-response effects, cumulative exposure, and timing of exposure, with particular focus on childhood and adolescence versus adulthood

6 Conclusion

This chapter has explored the intricate and multidimensional relationship between nature and human health, focusing the growing need to re-establish meaningful human-nature connections in the face of rapid urbanization. It has demonstrated that natural environments offer profound physiological, psychological, and social benefits from cardiovascular health and stress reduction to cognitive restoration and enhanced community cohesion. Despite these benefits, unequal access to nature, compounded by socioeconomic and spatial inequities, continues to undermine urban health and wellbeing. To address these challenges, we propose a conceptual framework (see Fig. 3) that integrates city infrastructure, natural systems, and human behavior as interconnected pillars of public health.

As illustrated in Fig. 3, the proposed conceptual framework underscores the interconnectedness of city infrastructure, natural systems, and human behavior in shaping public health and wellbeing. Within this model, urban green spaces including parks, green roofs, forests, and community gardens serve as pivotal links between nature and

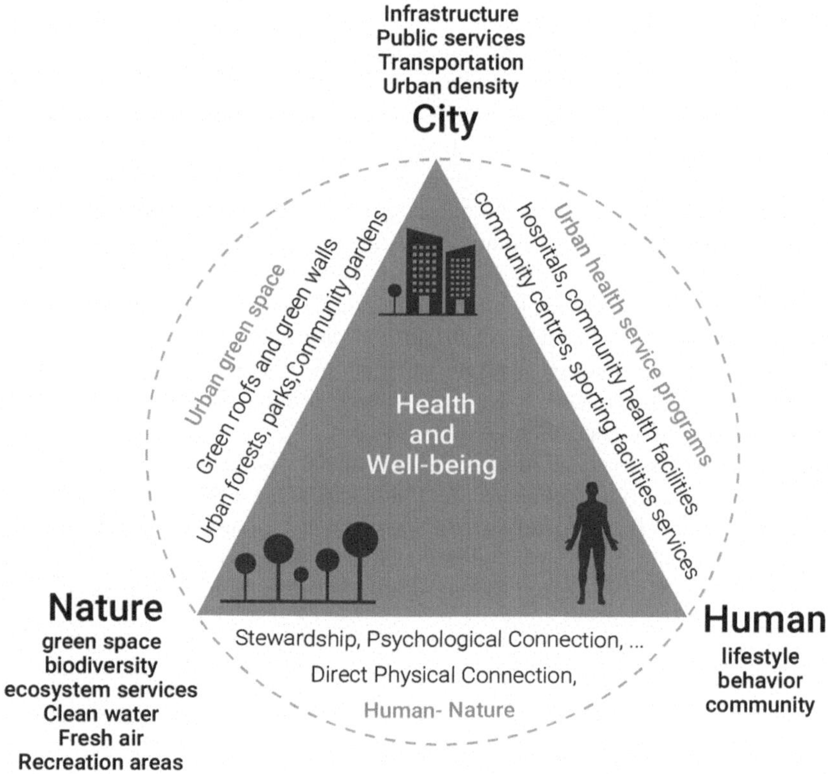

Fig. 3 Linking nature and human health in urban environments

daily urban life, offering essential ecosystem services such as clean air, biodiversity, and recreation. Simultaneously, access to health service programs and supportive public infrastructure facilitates healthy behaviors and community cohesion. The framework emphasizes that direct physical and psychological connections to nature, alongside responsible stewardship, are fundamental for realizing the full spectrum of physiological, psychological, and social benefits. By integrating these multidimensional elements, the framework addresses disparities in nature access and calls for urban planning approaches that holistically promote human health in rapidly urbanizing environments. At the core of this model is health and wellbeing, supported by three domains:

- Urban systems (e.g., infrastructure, services, density),
- Natural environments (e.g., green spaces, biodiversity, clean air, and water), and.
- Human dimensions (e.g., lifestyle, community, psychological connection).

These domains interact through mechanisms such as stewardship, physical activity, psychological restoration, and access to supportive services. This integrated approach highlights the importance of designing cities that not only incorporate green infrastructure and health services, but also foster human-nature connectedness as a form of preventive public health.

To fully harness the health-promoting potential of natural environments, this framework must integrate nature into the core of planning, policy, and design. Additionally, other approaches such as nature-based solutions, green infrastructure, and biophilic urbanism should provide promising pathways for creating resilient, health-supportive cities. However, realizing this vision demands more than just physical greening it requires fostering emotional, cognitive, and cultural connections to the natural world. Going forward, planners, designers, and policymakers must operationalize this framework by embedding it into urban strategies that promote equitable access to nature, especially for vulnerable populations. By bridging the physical, ecological, and social dimensions of urban life, cities can become more inclusive, resilient, and health-promoting. Reconnecting people with nature is not just an environmental or aesthetic concern it is a foundational strategy for improving collective wellbeing and achieving sustainable urban futures.

Future research must also address existing methodological gaps, ensure inclusivity, and provide robust evidence to guide equitable urban interventions. By bridging the divide between urban residents and the natural environment, cities can become not only more sustainable and livable but also more just and health-promoting. Ultimately, investing in nature is an investment in collective wellbeing and planetary health.

References

Anguelovski I, Brand AL, Connolly JJT et al (2020) Expanding the boundaries of justice in urban greening scholarship: toward an emancipatory, antisubordination, intersectional, and relational approach. Ann Am Assoc Geogr 110:1743–1769

Aras SG, Runyon JR, Kazman JB et al (2024) Is greener better? Quantifying the impact of a nature walk on stress reduction using HRV and saliva cortisol biomarkers. Int J Environ Res Public Health 21:1491

Barrable A, Booth D (2022) Disconnected: what can we learn from individuals with very low nature connection? Int J Environ Res Public Health 19:8021

Barragan-Jason G, Loreau M, de Mazancourt C et al (2023) Psychological and physical connections with nature improve both human Well-being and nature conservation: a systematic review of meta-analyses. Biol Conserv 277:109842

Bashan D, Colléony A, Shwartz A (2021) Urban versus rural? The effects of residential status on species identification skills and connection to nature. People Nat 3:347–358

Beery T, Stahl Olafsson A, Gentin S et al (2023) Disconnection from nature: expanding our understanding of human—nature relations. People Nat 5:470–488

Bell S, Fleming LE, Grellier J et al (2021) Urban blue spaces: planning and design for water, health and Well-being. Routledge

Colléony A, Levontin L, Shwartz A (2020) Promoting meaningful and positive nature interactions for visitors to green spaces. Conserv Biol 34:1373–1382

Dzhambov AM, Lercher P, Markevych I et al (2022) Natural and built environments and blood pressure of Alpine schoolchildren. Environ Res 204:111925

Edwards RC, Larson BMH, Burdsey D (2023) What limits Muslim communities' access to nature? Barriers and opportunities in the United Kingdom. Environ Plan E Nat Sp 6:880–900

Farhadi R, Soltanifard H, Alizadeh B (2025) Enhancing climate resilience through urban landscape design: strategies for green infrastructure and adaptation. In: Urban Climate and Urban Design Springer:117–128

Farhadi R, Ziari K, Bakhshayeshi S, Noroozi M, Nasehi S, Askarizad R (2024) Sustainable Urban Design for stormwater management: concepts and reviews. In: Cheshmehzangi A, Sedrez M, Flynn A (eds) Rethinking stormwater management through sustainable Urban Design. Urban Sustainability. Springer, Singapore. https://doi.org/10.1007/978-981-97-4924-9_9

Geary RS, Thompson DA, Garrett JK et al (2023) Green-blue space exposure changes and impact on individual-level Well-being and mental health: a population-wide dynamic longitudinal panel study with linked survey data. Public Heal Res 11

Geneshka M, Coventry P, Cruz J, Gilbody S (2021) Relationship between green and blue spaces with mental and physical health: a systematic review of longitudinal observational studies. Int J Environ Res Public Health 18:9010

Gómez-Baggethun E, Barton DN (2013) Classifying and valuing ecosystem services for urban planning. Ecol Econ 86:235–245

Hansen MM, Jones R, Tocchini K (2017) Shinrin-yoku (forest bathing) and nature therapy: a state-of-the-art review. Int J Environ Res Public Health 14:851

Hartig T, Mitchell R, De Vries S, Frumkin H (2014) Nature and health. Annu Rev Public Health 35:207–228

Hedblom M, Mårtensson F, Sang ÅO et al (2024) Play biotopes put into practice—creating synergies between children and nature. People Nat 6:2046–2059

Huang Y, Napawan NC (2021) "Separate but equal?" understanding gender differences in urban park usage and its implications for gender-inclusive design. Landsc J 40:1–16

Ives CD, Abson DJ, Von Wehrden H et al (2018) Reconnecting with nature for sustainability. Sustain Sci 13:1389–1397

Ives C, Oke C, Cooke B, et al (2014) Planning for green open space in urbanising landscapes

Iyer HS, James P, Valeri L et al (2020) The association between neighborhood greenness and incidence of lethal prostate cancer: a prospective cohort study. Environ Epidemiol 4:e091

Jimenez MP, DeVille NV, Elliott EG et al (2021) Associations between nature exposure and health: a review of the evidence. Int J Environ Res Public Health 18:4790
Kaplan R (1998) With people in mind: design and management of everyday nature
Keeler BL, Hamel P, McPhearson T et al (2019) Social-ecological and technological factors moderate the value of urban nature. Nat Sustain 2:29–38
Keniger LE, Gaston KJ, Irvine KN, Fuller RA (2013) What are the benefits of interacting with nature? Int J Environ Res Public Health 10:913–935
Koselka EPD, Weidner LC, Minasov A et al (2019) Walking green: developing an evidence base for nature prescriptions. Int J Environ Res Public Health 16:4338
Kruize H, van Kamp I, van den Berg M et al (2020) Exploring mechanisms underlying the relationship between the natural outdoor environment and health and Well-being—results from the PHENOTYPE project. Environ Int 134:105173
Lehmann S (2023) Reconnecting with nature: developing urban spaces in the age of climate change. Emerald Open Res 1
LoTemplio S, Bettmann JE, Scott E, Blumenthal E (2023) Do mental health changes in nature co-occur with changes in heartrate variability and executive functioning? A systematic review. Curr Environ Heal Reports 10:278–290
Lu M, Murphy M, Kim A et al (2024) The relationship between natural environments and obesity: a systematic review. Int J Environ Health Res:1–21
Mattijssen TJM, Buijs AAE, Elands BHM et al (2019) The transformative potential of active citizenship: understanding changes in local governance practices. Sustainability 11:5781
Newman P (2014) Biophilic urbanism: a case study on Singapore. Aust Plan 51:47–65
Newman P, Beatley T, Boyer H (2017) Resilient cities: Overcoming fossil fuel dependence. Island Press
Nieuwenhuijsen MJ, Kruize H, Gidlow C et al (2014) Positive health effects of the natural outdoor environment in typical populations in different regions in Europe (PHENOTYPE): a study programme protocol. BMJ Open 4:e004951
Polko P, Kimic K (2022) Gender as a factor differentiating the perceptions of safety in urban parks. Ain Shams Eng J 13:101608
Quah A, Haughton M, Kotera Y (2024) Investigating the relationship between nature exposure, nature connectedness and wellbeing in individuals with cancer. Int J Spa Wellness 7:250–276
Riechers M, Martin-López B, Fischer J (2022) Human—nature connectedness and other relational values are negatively affected by landscape simplification: insights from Lower Saxony, Germany. Sustain Sci 17:865–877
Sarkar S, Cottineau-Mugadza C, Wolf LJ (2024) Spatial inequalities and cities: a review. Environ Plan B Urban Anal City Sci 51:1391–1407
Shahamati S (2020) Assessment of landscape quality based on the perception of people: study of two parks in Newcastle upon Tyne. Eur J Environ Sci 10:65–75
Song C, Ikei H, Miyazaki Y (2016) Physiological effects of nature therapy: a review of the research in Japan. Int J Environ Res Public Health 13:781
Spotswood E, Benjamin M, Stoneburner L, et al (2021) Who has nature during the pandemic? COVID-19 cases track widespread inequity in nature access across the United States
Teixeira A, Gabriel R, Quaresma L et al (2021) Obesity and natural spaces in adults and older people: a systematic review. J Phys Act Health 18:714–727
Thomson G, Newman P (2021) Green infrastructure and biophilic urbanism as tools for integrating resource efficient and ecological cities. Urban Plan 6:75–88
Wang Y, Timms F, Li J, Yu H (2024) Benefits of nature-based intervention in combating the impact of urbanization on psychopathology in industrialized societies. Curr Opin Psychiatry 37:202–211
Warburton DER, Bredin SSD (2017) Health benefits of physical activity: a systematic review of current systematic reviews. Curr Opin Cardiol 32:541–556
Wen Y, Yan Q, Pan Y et al (2019) Medical empirical research on forest bathing (Shinrin-yoku): a systematic review. Environ Health Prev Med 24:1–21

Wicks C, Barton J, Orbell S, Andrews L (2022) Psychological benefits of outdoor physical activity in natural versus urban environments: a systematic review and meta-analysis of experimental studies. Appl Psychol Heal Well-Being 14:1037–1061

Williams TG, Logan TM, Zuo CT et al (2020) Parks and safety: a comparative study of green space access and inequity in five US cities. Landsc Urban Plan 201:103841

Wüstemann H, Kalisch D, Kolbe J (2017) Access to urban green space and environmental inequalities in Germany. Landsc Urban Plan 164:124–131

Yao W, Zhang X, Gong Q (2021) The effect of exposure to the natural environment on stress reduction: a meta-analysis. Urban For & urban Green 57:126932

Ye X, Dou Z, Jiang M et al (2023) Effects of Linpan nature therapy on health benefits in older women with and without hypertension. Front Public Heal 11:1208481

Zelenski J, Warber S, Robinson JM, et al (2023) Nature connection: providing a pathway from personal to planetary health

Zhong W, Schröder T, Bekkering J (2024) Implementing biophilic design in architecture through three-dimensional green spaces: guidelines for building technologies, plant selection, and maintenance. J Build Eng 92:109648

Zingoni A, Fionda C, Borrelli C et al (2017) Natural killer cell response to chemotherapy-stressed cancer cells: role in tumor immunosurveillance. Front Immunol 8:1194

Tackling the Challenges in Greening Our Cities for Health and Wellbeing

Carmel Williams, Yonatal Tefera, John Kandulu, Melanie Lowe, Thami Croeser, Jacinta Humphrey, and Veronica Soebarto

Abstract Research showing the impacts of green space and ecosystem health on physical and mental health has significantly increased over recent years. While the specific mechanisms continue to be explored, the evidence is clear: the quantity and quality of green space, particularly those with high biodiversity and substantial tree canopy, are key to promoting health and livability in cities. Green space provides important habitat for native wildlife and mitigates the risks associated with a warming climate and rapid urbanisation. It reduces the health effects of noise and air pollution, improves social and economic wellbeing, encourages physical activity, fosters social

C. Williams (✉) · Y. Tefera
Centre for Health in All Policies Research Translation, South Australian Health and Medical Research Institute, Adelaide, SA, Australia
e-mail: Carmel.Williams@adelaide.edu.au

Y. Tefera
e-mail: Yonatal.Tefera@healthtranslationsa.org.au

C. Williams
School of Public Health, The University of Adelaide, Adelaide, SA, Australia

J. Kandulu
College of Business, Government and Law, Flinders University, Adelaide, SA, Australia
e-mail: John.Kandulu@flinders.edu.au

M. Lowe
Centre for Urban Research, School of Global, Urban and Social Studies, RMIT University, Melbourne, VIC, Australia
e-mail: Melanie.Lowe@rmit.edu.au

T. Croeser · J. Humphrey
ICON Science, Centre for Urban Research, School of Global, Urban and Social Studies, RMIT University, Melbourne, VIC, Australia
e-mail: Thami.Croeser@rmit.edu.au

J. Humphrey
e-mail: Jacinta.Humphrey@rmit.edu.au

V. Soebarto
School of Architecture and Civil Engineering, The University of Adelaide, Adelaide, SA, Australia
e-mail: Veronica.Soebarto@adelaide.edu.au

A. Cheshmehzangi et al. (eds.), *Nature in Cities, Nurturing Cities*, Urban Sustainability,
https://doi.org/10.1007/978-981-95-5487-4_3

connections and contact with nature, reduces stress, boosts mental health, and reduces mortality. Despite these known benefits, the design and delivery of green spaces that promote health and wellbeing in an equitable manner remains a challenge in rapidly growing and densifying cities. The need for higher-density, more compact urban development to accommodate population growth often comes at the expense of public and private urban green space. While there are growing greening policy commitments from governments, a concerning decline in these natural assets persists in many cities internationally, including in Australia. Several identified challenges that impact the greening of cities include: • The lack of economic incentives to prioritise green spaces over other land uses on both public and private land. • The lack of joined-up policy action and competing policy priorities with insufficient coordination across sectors and levels of government. • Inequitable distribution of green spaces across communities, contributing to health inequities. This chapter will explore these challenges by unpacking their underlying drivers, both internationally and within the Australian context. It also identifies potential levers and actionable strategies that can help retain, enhance, and deliver new urban green spaces that promote health and wellbeing.

Keywords Green equity · Health and wellbeing · Public and private green spaces · Economic value · Public policy · Multisectoral action

1 Introduction

Urbanisation is a transformative global phenomenon, with an accelerated shift of populations from rural to urban areas in recent decades. Currently, around 56% of the world's population resides in urban centres (Kuddus et al. 2020; The World Bank 2023), and more than two-thirds of the global population is projected to live in cities by 2050 (The World Bank 2023). This unprecedented urban growth has profound risks for urban equity, health, and wellbeing (Ramirez-Rubio et al. 2019). Simultaneously, there are opportunities to enhance wellbeing through innovative urban planning, creating human- and nature-sensitive neighbourhoods and communities (Bai et al. 2012).

The interconnected planetary crises of climate change, biodiversity loss and pollution, and escalating associated health issues demand global-scale action from public, private, and civil society actors. However, there is also an urgent need for national, regional, and local governments to address the impacts of these crises at the local level—where communities often feel their effects most directly. For urban populations, green space holds promise as a strategy to reduce climate change impacts, promote population health, support biodiversity, and contribute to greenhouse gas emissions reductions.

While there is no single agreed definition of urban green space, the term has been used to refer to both natural areas and any vegetated spaces in cities, including individual trees (Taylor & Hochuli 2017). Hereafter, we consider urban green space to be formally managed, vegetated spaces in urban areas including parks, streetscapes, front and back yards, sport fields, and school grounds. Urban green space is crucial as it provides amenity for people, supports native species, and facilitates a range of ecosystem services that help make cities healthier and more liveable.

Despite a growing body of evidence of the health and environmental benefits of urban green space and commitments from governments to expand green space and tree canopy cover, many cities worldwide, including in Australia, are experiencing a decline in both. This loss has significant consequences for both population health and wellbeing and the environment, particularly by amplifying the risks associated with a changing climate.

This chapter will explore the drivers of this decline and propose practical strategies to halt and reverse the trend—ultimately leading to improved health and environmental outcomes. As the focus of this chapter is on urban environments, the term "urban green space" will be simplified to "green space".

2 Understanding the Benefits of Green Space

Nature is increasingly recognised as an upstream strategy for health promotion and disease prevention, with growing interest from research, clinical, and policy sectors. Approaching health within an ecological paradigm highlights how green space can be a low-cost, accessible, and equitable resource for both preventative and restorative public health interventions (Maller et al. 2006). Exposure to green space can be either incidental, occurring during daily activities such as walking or cycling to local shops or schools or intentional, where individuals are outdoors to partake in gardening or bushwalking; or recreational, where they seek out green environments for rest, recovery, or relaxation under a tree.

2.1 Health and Wellbeing Benefits

A growing body of evidence supports the positive impact of green space exposure on both physical and mental health, though the specific mechanisms, optimal "dosage", and frequency remain under investigation. Wolf et al. (2020) categorise the benefits of urban trees and green infrastructure into three areas: capacity restoration (e.g., improved mental health, reduced stress, better clinical outcomes); capacity building (e.g., enhanced birth outcomes, increased physical activity); and harm reduction (e.g., mitigation of air pollution, UV exposure, and heat stress).

A systematic review (Yang et al. 2021) linked green space exposure to improvements across more than 100 health outcomes. Consistent positive associations were found with reduced all-cause and cardiovascular mortality (Rojas-Rueda et al. 2019; Yuan et al. 2021), lower cardiometabolic risk (Luo et al. 2020), better mental wellbeing (de Keijzer et al. 2016), reduced risk of low birthweight (Hu et al. 2021), decreased physical inactivity (Lambert et al. 2019), improved sleep (Shin et al. 2020), and enhanced immune function (Andersen et al. 2021). However, findings were less consistent for outcomes such as cancer, asthma, chronic respiratory diseases, preterm birth, small-for-gestational-age, and academic performance (Yang et al. 2021).

A recent narrative review found evidence that greater nature exposure is associated with improved cognitive performance, beneficial changes in brain function, lower blood pressure, better sleep, and reduced incidence of cardiovascular diseases (Jimenez et al. 2021). Many studies report higher physical activity levels in greener neighbourhoods, which likely mediates some of the health advantages (Almanza et al. 2012; Coombes et al. 2010; Mytton et al. 2012). Epidemiological and experimental studies have documented lower stress, better mood, and enhanced cognitive function after contact with nature (Mygind et al. 2021; Zhang et al. 2020).

2.2 Mechanisms Linking Green Space to Health and Wellbeing

While the benefits are well-documented, a key question in the current research is how green space exposure leads to improved health and wellbeing outcomes. According to Jimenez et al. (2021), two foundational theories from environmental psychology offer useful insights. Attention Restoration Theory (ART) suggests that urban environments place high demands on directed attention, leading to mental fatigue. In contrast, natural environments engage involuntary attention in a soft and effortless way, helping to restore cognitive capacity. Stress Reduction Theory (SRT) proposes that humans have an innate connection to nature, and exposure to natural settings activates the parasympathetic nervous system, promoting relaxation and reducing stress. Empirical studies support both theories, with individuals frequently reporting improved mood, reduced stress, and cognitive renewal after spending time in parks or green landscapes.

Beyond these cognitive and emotional pathways, researchers have identified multiple synergistic mechanisms through which green space influences health including sensory-perceptual and immunological processes, air quality, physical activity, stress, and social integration (Barton and Rogerson 2017). Green settings are attractive places for physical activity and play (walking, cycling, sports), thereby raising overall fitness and mobility (Jimenez et al. 2021), and reducing obesity and metabolic risk (Akaraci et al. 2022). However, multilevel analysis suggests that physical activity alone does not fully explain the relationship between green space and health (Maas et al. 2008). As social gathering places, parks, community gardens, and nature-based events create environments where people can interact and connect. Observational data links park proximity and quality to stronger neighbourhood social ties and perceived social support (Cui et al. 2024), both of which are known to contribute to better health outcomes (Jimenez et al. 2021). Such social and behavioral pathways further reduce loneliness, support healthy ageing, and promote health and resilience (Astell-Burt et al. 2022, 2024; John et al. 2023).

Green space enhances broader environmental quality, offering a range of ecosystem services that translate into important public health benefits (Wolch et al. 2014). Green infrastructure serves a critical role in climate adaptation by buffering cities against extreme weather events, such as heatwaves and heavy rainfall, and strengthening urban resilience in the face of climate change (Pihkala 2020). Urban vegetation helps to mitigate environmental hazards by filtering air pollutants such as particulate matter and ozone precursors, regulating temperatures, and reducing the urban heat island effect (Islam et al. 2024; Knight et al. 2021). These latter microclimate benefits can help lower the incidence of heat-related illnesses and reduce energy demand for cooling, thereby decreasing pollution and emissions. Cleaner air and reduced temperatures can also lower the risk of cardiovascular and respiratory diseases, particularly in vulnerable populations. In addition, trees and green infrastructure help regulate and purify water, intercept stormwater, and improve drainage systems, reducing the risks of flooding and enhancing water quality (Yang et al. 2015; Zimmermann et al. 2016). Green space also supports biodiversity by providing habitats for pollinators and other wildlife (Lepczyk et al. 2017) and may enrich the urban microbial environment in ways that benefit immune function (Zhang et al. 2024). The inclusion of water features and diverse plant life in green spaces, sometimes referred to as "blue-green infrastructure", can enhance aesthetic appeal and provide restorative experiences that support mental health.

Further interdisciplinary research is needed to disentangle how various pathways interact to produce the observed physical, mental, and social health benefits of green space exposure (Hunter et al. 2023). The mechanisms are likely to be complex, multifactorial, and context dependent.

2.3 *Dose–Response Relationship*

While many studies demonstrate the health benefits of green space exposure, relatively few have investigated the dose–response relationship, that is, how much time or how often people need to engage with nature to experience measurable health improvements. Understanding this relationship is essential for developing practical, evidence-based health promotion recommendations.

Shanahan et al. (2016) used a "nature dose" framework and found that longer visits to green space were associated with significantly lower rates of depression and high blood pressure. Meanwhile, more frequent visits were linked to higher levels of physical activity and stronger social cohesion. Their analysis suggested that just 30 min of green space exposure per week could reduce the population prevalence of depression by 7% and hypertension by 9%, a substantial impact in public health terms. Similarly, White et al. (2019) showed that spending at least 120 min per week in nature was associated with better self-reported health and wellbeing. The benefits peaked between 200 and 300 min per week, with no additional gains beyond that threshold. Experimental studies support these findings; for example, Bratman et al. (2015) reported that a single 90-min walk in a natural environment significantly reduced self-reported rumination and neural activity in the subgenual prefrontal cortex, a brain region linked to depressive thinking, compared to a similar walk in an urban setting.

Overall, despite these promising insights, most of the existing evidence is drawn from cross-sectional studies, which limits causal inference. More robust research using longitudinal designs or experimental interventions is still needed to determine the optimal "dose" of nature for different populations and health outcomes (Yang et al. 2021).

3 Quality, Quantity, and Distribution of Green Space

3.1 Green Space Quality

Higher quality green spaces may have greater benefits for human health, mental wellbeing, and social connection. Some studies have suggested that green space quality may in fact be more important than its distribution (van Dillen et al. 2012; Zhang et al. 2017), particularly for enhancing mental health outcomes (Francis et al. 2012). There are many different components to green space quality and the concept of a "high quality" green space varies according to discipline. Some established components of green space quality include quantity, size, amenities or facilities, and biodiversity.

3.1.1 Quantity

Under-provision of green space can result in crowding. Green space crowding has been shown to reduce its benefits to mood, particularly for women (Honey-Rosés and Zapata 2023; Koohsari et al. 2024). Crowding can lead to avoidance of green space (Arnberger 2012; Otto et al. 2024), and may result in conflicts over its availability for socially valuable but space-consuming sporting uses, both formal and informal (Jeanes et al. 2019), and limit opportunities for children's play (Lu and van Ameijde 2025).

Close attention to ensuring adequate green space is especially important in urban settings where infill development (i.e., densification) acts to rapidly increase the local population while reducing private green space. While urban density offers many benefits (e.g., walkability, sociability), unless well managed, it places pressure on existing public infrastructure and facilities, including green spaces. In the context of densification, conserving and increasing public green spaces even as land use competition increases is critical for sustaining and enhancing residents' health (He et al. 2022; Koohsari et al. 2024; Zhang et al. 2023).

The amount of tree canopy cover in a neighbourhood is also important, given evidence of its benefits for urban cooling and health. 30% tree canopy cover is a common policy target internationally. Australian evidence suggests that living in areas with at least 30% tree canopy cover may reduce psychological distress and risk of dementia, loneliness, and cardiometabolic illness (Astell-Burt and Feng 2019, 2020; Astell-Burt et al. 2022), with significant economic benefits to society (Feng et al. 2024).

3.1.2 Size

Green spaces vary in size, both within and between different urban contexts. Larger green spaces (e.g., at least 1 ha in size) encourage recreational physical activity (e.g., walking, running, bike riding), which enable visitors to achieve recommended levels of exercise (Davern et al. 2024; Giles-Corti et al. 2005; Jansen et al. 2017). Large, continuous green spaces are also important areas for native wildlife, particularly more sensitive species that require relatively undisturbed, intact patches of habitat (Dale 2018; Humphrey et al. 2024). Relative to small spaces, larger green spaces may also have greater cooling effects (Jaganmohan et al. 2016).

While smaller green spaces are useful for passive recreation, they typically do not provide the same level of health and wellbeing benefits for people (Giles-Corti et al. 2005; Holt et al. 2019). For this reason, very small green spaces are often excluded from distribution and access metrics (e.g., the World Health Organization excludes all green spaces <0.5 ha (World Health Organization 2017)).

3.1.3 Amenities or Facilities

Green space quality is also determined by the presence, quality, and suitability of its amenities and facilities. Community needs vary according to population demographics, but in general, green space can be considered higher quality when it caters to a broader range of activities, ages, genders, abilities, and ethnic groups (Davern et al. 2024). Some key amenities that may increase the quality of green space include dog exercise areas, playgrounds, shelters, picnic areas, seating, drinking fountains, BBQs, rubbish bins, lighting, walking or running paths, public toilets, informal sport areas (e.g., basketball half courts, BMX tracks/or jumps), formal sport areas (e.g., football fields, tennis courts), directional signage, car parks, bike racks, bird hides, nature trails, and waterbodies. The level or frequency of maintenance is also important (e.g., mowing, pruning, irrigation, cleaning) as this will influence the quality and useability of facilities over time (Creed and Carvalho 2024).

3.1.4 Biodiversity

One of the most critical aspects of green space quality is biodiversity (Carrus et al. 2015), that is, the variety of living organisms that occur in an urban green space. The literature suggests that there are significant additional health and wellbeing benefits when green spaces are *biodiverse*, rather than just "green" (Wood et al. 2018). For instance, it is entirely possible to fill a green space with non-native tree species that may be aesthetically pleasing but provide very little in the way of habitat or food resources for wildlife. Research suggests there can be positive associations between green space species diversity (such as the diversity of flowers or birds) and mental wellbeing (Cameron et al. 2020; Houlden et al. 2021; Wood et al. 2018), as well as ecosystem diversity and immune system function (Aerts et al. 2018). Wildlife encounters are particularly beneficial for green space users, with birdwatching known to reduce feelings of anxiety and stress while promoting a sense of immersion or attention, and enhancing opportunities for social connection (Andrews et al. 2024; Bell et al. 2018; Peterson et al. 2024). In general, green space is more likely to be biodiverse when it is large, contains a diversity of native vegetation (including groundcover plants, mid-storey shrubs, and canopy trees), incorporates waterbodies (such as wetlands, lakes, and rivers), and connects to other nearby green spaces or remnant natural ecosystems (Callaghan et al. 2018; Chamberlain et al. 2007; de Toledo et al. 2012; Ferenc et al. 2014; Ikin et al. 2015; Mata et al. 2021; Threlfall et al. 2017).

3.2 *Green Space Distribution*

Green space is often unequally distributed across cities and communities (Astell-Burt et al. 2014; Hsu et al. 2022; Mushangwe et al. 2021; Rosario et al. 2025). While the health and wellbeing benefits of nature are well established, not all communities have equal access to high-quality green environments. Access is often patterned along socio-economic lines, raising concerns about environmental justice and equity (Kronenberg et al. 2020).

3.2.1 Equality and Equity

Lower socio-economic areas with ethnically diverse populations often experience reduced access to green space compared with more affluent suburbs. This trend has been identified in parts of Europe (Hoffimann et al. 2017; Schüle et al. 2017), the USA (Dai 2011; Liu et al. 2021), and Australia (Astell-Burt et al. 2014; Davern et al. 2024; Jamalishahni et al. 2023). Further, there can also be inequality in the distribution of high-quality green spaces. Not all green spaces are of equal value for people or nature, and higher quality green space tends to occur in wealthier neighbourhoods (de Vries et al. 2020; European Environment Agency 2022; Viinikka et al. 2023; Wright

Wendel et al. 2012). These issues are of particular concern, as research indicates that the health and wellbeing benefits afforded by green space are often strongest among members of lower socio-economic groups and ethnic minorities (Mitchell et al. 2015; Rigolon et al. 2021).

3.2.2 Measuring Distribution of Green Space

The distribution or accessibility of green space in an urban context is often measured by the distance people must walk from home to reach their nearest green space. Good access and closer proximity to green space are associated with higher physical activity levels and more frequent visits (Coombes et al. 2010; Kaczynski et al. 2009), resulting in enhanced health outcomes for residents (Irvine et al. 2013; Konijnendijk 2023; Toftager et al. 2011). For example, the World Health Organization recommends that all urban residents should be able to access an area of green space (of at least 0.5–1 ha) within 300 m of their home, or approximately a five-minute walk (World Health Organization 2017). This guideline has recently been incorporated into the "3–30-300 rule", which recommends all households, schools, and workplaces have a view of at least three trees, a minimum of 30% canopy tree cover (at the neighbourhood scale), and a maximum of a 300 m walk to the nearest green space (Konijnendijk 2023). A recent international study found that even in very high-profile cities such as Amsterdam and New York City, full achievement of this standard was often low (Croeser et al. 2024).

The distribution of green space varies considerably in Australian cities, with large capital cities enjoying better access to green space when compared to smaller, regional cities (Australian Urban Observatory 2025). For example, in Australia's three largest cities, Sydney, Melbourne, and Brisbane, 35–46% of homes were located within 400 m of a 1.5 ha green space (Australian Urban Observatory 2025). In contrast, only 25% of homes meet the mark in smaller urban centres like Launceston and Toowoomba (Australian Urban Observatory 2025).

In an important progression from this type of benchmark, cumulative opportunity metrics, which consider all nearby green spaces rather than just the single closest site, may offer a more complete understanding of the availability of green space opportunities for residents (Ekkel and de Vries 2017).

4 Understanding Challenges of Implementing Green Space

Despite its clear health and environmental benefits, efforts to retain, improve, and expand urban green spaces face many challenges, particularly in Australia. For example, the Greater Adelaide Metropolitan area faces acute challenges as population growth and urban densification intensify competition for limited land. Tree canopy cover in Greater Adelaide stands at just 17%—well below the 30% target outlined in the 30-Year Plan for Greater Adelaide—with ongoing tree removals particularly prevalent in high-growth residential areas (Green Adelaide 2024). Despite considerable public investment in tree planting, the city's overall urban tree canopy continues to decline, with losses on private land emerging as a major driver. Notably, in western Adelaide, private backyards occupy about 20% of urban land but account for over 40% of total tree cover. These losses are driven by infill housing and subdivision: large lots are split and mature trees are cleared for new driveways and homes. Upgrades to public space (e.g., streetscape greening and park expansion) to compensate losses of private green spaces are rarely integrated into densification initiatives.

This section explores several challenges for implementing, maintaining, and expanding green space, drawing on examples from Australian cities. These issues include: (1) undervaluation of green space in urban planning; (2) loss of private green space and street trees associated with urban densification; (3) major clearing for new infrastructure and civic buildings; (4) vegetation clearing to create new housing; and (5) excessive tree removal. In many municipalities in Australian cities, these challenges have resulted in declining tree canopy cover and/or open space availability per capita.

4.1 Undervaluation of Green Space

Green space remains undervalued in current urban planning processes that shape urban environments. Failure to recognise green (and blue) space as essential urban infrastructures limits their integration into development frameworks, missing opportunities to deliver more resilient and health-promoting urban environments. More specifically, financial models used in planning decisions often fail to reflect the true economic value of nature and instead treat trees and green infrastructure as liabilities (Tefera et al. 2023). This mischaracterisation may be contributing to the continued decline of green space coverage across many cities, including in Australia.

A persistent research and policy gap in urban green space management is the inadequate valuation of the full suite of benefits provided by urban green space. Traditional cost–benefit analysis (CBA) frameworks often overlook indirect and intangible ecosystem services—such as improved biodiversity and mental health—particularly those delivered by trees on private land. Methodological challenges, including difficulties in monetising non-market values and inconsistencies in valuation techniques, further constrain comprehensive decision-making (Teotónio et al. 2021).

This complexity is compounded by the contextual variability of benefits and the sensitivity of CBA outcomes to uncertainties in data and modelling. There is a lack of context-specific data to support the development of appropriate CBA scope, establish baselines, and evaluate alternative greenspace scenarios that align with or challenge regional strategic goals. In particular, limited data on intangible outcomes such as mental wellbeing and noise reduction makes it difficult to build a compelling business case for investment in both public and private green space.

Systemic undervaluation leads to ineffective policy interventions (Chelli et al. 2025). Disincentive measures, such as fines for unauthorised tree removal, are often set without reference to the true social and environmental value of trees, rendering them ineffective deterrents. For example, in Adelaide, the permit fee to remove a regulated tree is AUD $1,000 and AUD $1,500 for a "significant" tree, while conservative estimates place the social benefit of a single tree at up to AUD $16,211 (Kandulu and Soebarto 2023). Similarly, incentive programs—such as subsidies or rate rebates for retaining trees—are frequently underfunded or poorly targeted and thus fail to capture and reward the substantial public benefits provided by mature trees (Ordóñez-Barona et al. 2021). Without stronger alignment of incentives, private owners and developers tend to favor clearing because it maximises buildable area and avoids upkeep, resulting in persistent downward pressure on private-land tree canopy.

4.2 *Densification in Established Urban Areas*

Infill development—or densification of housing—is proceeding at a rapid pace in many cities internationally. Development pressures on green space can be particularly acute in the most rapidly growing cities in low- and middle-income countries, but can also be an issue in high-income countries, including several major Australian cities. Traditional models of infill have resulted in the loss of garden areas and tree canopy on private property, making way for large structures that cover most or all of the property. Since 2005, the average block size for new housing builds in Australian cities has declined by ~22%, yet the average floor area of new homes has increased (Australian Bureau of Statistics 2020). Public street trees can also be removed during development to facilitate site access for heavy machinery during construction, for new driveways, or to simplify underground utility provision (e.g., pipes, cables). The scale of tree loss during periods of rapid construction in established urban areas can be so great that even well-resourced municipal tree planting programs can make little or no overall progress (Croeser et al. 2024).

4.3 *Major Infrastructure*

Major infrastructure can produce very large tree losses, as well as the use of green open spaces as sites for new buildings, or as footings for large transport overpasses. Examples of this include a recent Melbourne highway widening project, which required the removal of 10,000 trees and 52 hectares of vegetation (North East Link 2025) and the decision by the Queensland State government to convert a large section of Brisbane's Victoria Park into an Olympic Stadium (ABC News 2025). Green spaces can also be impacted by infrastructure without being directly developed. Many parks suffer from road noise, for example, which limits their value as relaxation spaces for local communities (Van Renterghem 2024).

4.4 Vegetation Clearing for Greenfield Development

Remnant vegetation can have considerable conservation value, but this does not always lead to its protection in the face of residential development pressures. In Melbourne, for example, only 1% of the critically endangered Western Plains Grasslands ecosystem remains due to its incremental destruction following expansion of the city's urban growth boundary. A 2020 report by the Victorian Auditor-General's office found that offset schemes had delivered less than 10% of the promised compensatory protected areas. The few successfully protected remnants are currently closed to community use and have not been adequately managed to limit the impacts of weeds and feral species (Victorian Auditor-General's Office 2020). Together this amounts to a very poor green space outcome for local residents. A more contemporary example of this kind of offset-driven development exists in Sydney, where 70,000 new houses are planned for the Cumberland Plain, an area of high-value habitat for the state's endangered population of koalas (NSW Department of Planning 2025).

Greenfield vegetation loss is exacerbated by the development practice of comprehensively clearing, scraping, and leveling residential land (effectively obliterating remnant local ecosystems) as part of pre-construction site preparation (Parsons et al. 2023; Russell 2021). While this clearing is generally followed by some level of park creation and street tree planting, new residents experience highly modified, immature green infrastructure, with little prospect of connecting with original natural ecosystems. Recently, increasingly densely packed greenfield housing layouts have aggravated this issue by reducing the size of backyards and street verges, sometimes to the point where tree planting becomes extremely difficult.

4.5 Tree Removal

Tree canopy in urban areas is also lost to storm damage, disease, exotic pest infestations, drought, vandalism, and shorter tree lifespans due to adverse growing conditions (e.g., constrained soil volumes) (Clark et al. 2020; Smith et al. 2019). Residents can demand tree removals due to disservices such as fallen leaves, perceived risks of limb drop, or root damage, and due to aesthetic preferences (Conway 2016; Kirkpatrick et al. 2012; Roman et al. 2021). Disservices (real and perceived) are correlated with tree size, which makes larger trees more likely to be subject to these kinds of complaints, especially in contexts where trees are heavily pruned or pollarded (Masini et al. 2023). Poor management of tree removal requests can be disastrous. In Adelaide, for example, an exemption on needing a permit to clear trees within 10 m of properties was only revised in 2024 after it was revealed the city was losing 75,000 trees annually (ABC News 2022; Local Government Association of South Australia 2024). Offsetting tree loss with new planting is problematic, due to the difficulty of establishing young trees in heavily urbanised environments, and the intense contestation of public space in these areas, with vehicles and utilities generally taking priority due to stronger legal rights and higher social and institutional legitimacy as users of public space (and indeed, much larger capital works budgets).

5 Overcoming Challenges in Urban Green Space Planning and Management

To address the above challenges, this section will discuss potential government policy responses and tangible strategies that can be adopted into the urban planning and development system to help retain, enhance, and deliver new urban green spaces that promote health and wellbeing.

In particular, this section will focus on three policy/strategy areas that could help halt the decline in green space across our cities:

- The development and application of economic models that account for the true economic value of trees and green space to our cities and communities;
- Increasing the capacity of different governments sectors/ministries to work together and develop collaborative and impactful joined-up public policy responses for green space provision; and
- Approaches to support equitable green space access.

5.1 Improved Economic Modelling

Tackling green space decline requires economic models that account for the full value of green spaces, particularly its long-term social returns and context-specific benefits. Omitting intangible values such as biodiversity support, mental health improvements, and urban cooling can lead to "under-valued and under-provided" green infrastructure (Chelli et al. 2025; Teotónio et al. 2021). Frameworks that integrate ecosystem-service valuation into multi-horizon planning help to reveal the true costs and benefits of improved green space management.

Some of the chapter authors have developed a context-sensitive valuation approach, building on the foundational tree valuation analysis by Kandulu and Soebarto (2023). This approach integrates improved economic valuation with collaborative engagement of policymakers and the creation of a context-aware decision-support tool. It was piloted with two councils in Adelaide, Australia with contrasting green space planning challenges—the City of Salisbury and the City of Unley (see Box 1). The approach involved four key elements:

1. Collaborative scenario development was undertaken through workshops with policymakers to co-develop green space futures, ensuring that the tool and its outputs were grounded in local realities and policy aspirations. This collaborative

process highlighted the importance of tailoring strategies to local pressures—such as affordable housing demand in Salisbury and urban infill in Unley—and of addressing both socio-economic and environmental barriers.
2. Co-development of a decision-support tool that integrated an ecosystem services framework, enabling enumeration and monetisation of a broad spectrum of benefits, including amenity, air pollution reduction, cooling, carbon sequestration, stormwater management, and energy savings. The i-Tree Eco model (Nowak 2021) underpinned benefit valuation, providing a scientifically rigorous basis for quantifying ecosystem services.
3. The decision support tool features an intuitive user interface that enables councils to simulate green space management outcomes over 5-, 30-, and 60-year horizons. Users adjust parameters (e.g., tree species, establishment, and maintenance costs) and immediately visualise the long-term financial, health, and environmental implications of different choices.
4. The tool calculates total costs, total benefits, net present value (NPV), and benefit–cost ratio (BCR) for each scenario, providing a clear, evidence-based foundation for strategic decision-making and resource allocation. Real-time visualisation capabilities enable stakeholders to track performance and adjust strategies as needed.

This integrated approach enables evidence-informed, context-sensitive decision-making for optimising the benefits of green space. By providing a scientifically rigorous framework for CBA and an accessible platform for scenario analysis, the tool empowers councils to make strategic choices supported by quantifiable data. This helps justify investments in urban green space by demonstrating long-term returns. For example, comprehensive valuation of a broad set of benefits of urban trees can inform effective management of trees, including on private land, by aligning policy incentives and deterrents with the full spectrum of social, environmental, and economic benefits that urban trees provide. Further, enabling planners to quantify not only the direct financial returns but also the indirect and intangible benefits allows for more accurate assessment of trade-offs associated with densification, infrastructure development, and vegetation clearing, and supports the design of targeted incentives and disincentives that reflect the true social and environmental value of urban green spaces.

Box 1: Case Studies from the City of Salisbury and the City of Unley in Adelaide, Australia

Case Study Insights from City of Salisbury

Salisbury's population is young, with more people in the 0–9 and 20–39 age groups, and a larger proportion of families with children. It is culturally diverse and has a higher proportion of low-income households, with greater housing affordability pressures than Unley. Salisbury faces significant constraints in managing tree canopy on private land due to limited jurisdiction, strong housing affordability pressures, and changing demographics. Strategies must be equity-aware, addressing both socio-economic and environmental barriers. Targeted engagement with culturally diverse and low-income communities, alongside investment in digital compliance systems, are important for cost-effective implementation. A 'one tree per allotment' rule fails to offset losses as mature trees are replaced with smaller species or removed entirely during infill development. Smaller private yards (averaging 300 m^2 in new subdivisions) leave minimal space for large-canopy trees, while state-mandated infill targets prioritise housing over greenspace.

Case Study Insights from City of Unley

Unley has a median age of 43, above-average household incomes, and almost half its residents possess a bachelor's degree or higher. It has a high concentration of medium and high-density housing with elevated rental costs. Unley is challenged by ongoing infill and subdivision, which reduces yard space and accelerates tree decline on private lots. Weak planning codes and limited incentives for private landowners exacerbate the problem. Minimum landscaping standards allow developers to clear pre-existing trees on subdivided lots. Private garden sizes have decreased, accelerating tree loss. Policy reform, strengthened enforcement, and market-based tools (such as tree bonds) are essential to reducing canopy loss. Unley needs revised planning codes mandating canopy retention on subdivided lots.

Participatory tools that incorporate local knowledge can bridge the gap between theory and practice. By involving planners and residents in scenario development, the resulting outputs are actionable rather than abstract. Even if contextual details differ, cities worldwide can adapt this approach, involving co-designing tools with local stakeholders, explicitly modelling non-market benefits, and comparing scenarios over long timeframes. In doing so, cities can make a stronger evidence-informed business case for improving the extent and quality of green space.

5.2 *Joined Up Policymaking: The Case for an Environment in All Policies Approach*

Complex societal problems—such as green space loss—require healthy public policy and joined-up government and community action (Williams et al. 2023). While protecting the environment is the primary responsibility of environmental agencies within government, decisions affecting urban environments are often made outside this sector. However, the highly siloed structure of government often limits the integration of environmental or health considerations into urban policy decisions. For instance, urban and transport planning have a key role in determining access to urban nature and its human and planetary health benefits. Despite this, collaboration among urban planners, transport authorities, environmental agencies, green space planners, and the public health sector remains inconsistent and limited. This may be contributing to the suboptimal green space planning outcomes outlined in Sect. 4.

While whole-of-government discussions often take place at the highest levels during planning and budget processes, these can be challenging due to competing departmental priorities. Even when high-level agreement is reached on environmental action such as green space or tree canopy policy, implementation is often hampered by the compartmentalised structure of government departments and limited collaboration at operational levels, making cross-sectoral coordination difficult. Delivering equitable green space access requires transformative frameworks that foster transparency, reciprocity, and cross-sectoral coordination throughout all levels and areas of government. Additionally, sustained resources and mechanisms are needed to maintain these cross-sectoral systems and drive the necessary planning changes to protect and expand green space access.

Several cross sectoral mechanisms have been proposed, developed, trialled, and in some cases, found to be successful. For example, the Health in All Policies (HiAP) approach (World Health Organization 2014) has been developed and implemented in various places, including South Australia, Finland, California, New Zealand, and Canada. HiAP has been developed for use by the health sector, providing structured advice on the whole-of-government mechanisms that need to be in place to enable joined up public policymaking that considers the health implications of all government policy (World Health Organization 2014). At its core, HiAP promotes strong, effective governance by fostering trust, collaboration, and mutual benefit across sectors to enhance health and wellbeing outcomes.

The Four Pillars Model of Health in All Policies (HiAP) (Fig. 1), established by the WHO in 2023 (World Health Organization 2023), is a flexible framework that can be adapted to various contexts, sectors, and challenges. This model offers practical guidance on essential components needed to establish and maintain collaboration for achieving health outcomes.

Given the environment and health co-benefits of green space, we propose adopting an Environment in All Policies (EiAP) 4 Pillars Model (Fig. 2), which adapts the HiAP model, to be relevant to and used by the environment sector, which has primary responsibility for green space. Building on the HiAP ensures that the same four pillars—the structures and mechanisms that enable joined-up government decision-making—are being pursued by both the health and environment sectors. The pillars focus on leadership for collaborative approaches; governance and accountability mechanisms; partnerships and joined-up ways of working; and resources and capabilities. The difference between the HiAP and the EiAP is the latter model explicitly considers human–environment interactions and synergies, which is highly relevant to the multisectoral challenge of green space planning and protection. The EiAP model emphasise the importance of publicly reporting the data on health and environment (under the Governance and Accountability pillar) and the need to build community knowledge to act and advocate for environment (under pillar of Leadership at All Levels). The EiAP model also emphasises the need to identify areas for collaboration (under the pillar of Ways of Working & Work Methods) as well as to strengthen knowledge and understanding of the synergies between the environment and health (under Resource, Financing and Capabilities pillar). For example, when considering new infrastructure developments on public land, an EiAP approach will ensure that environmental implications are factored into the policy decisions about whether and how to protect existing green space, with co-benefits for human health.

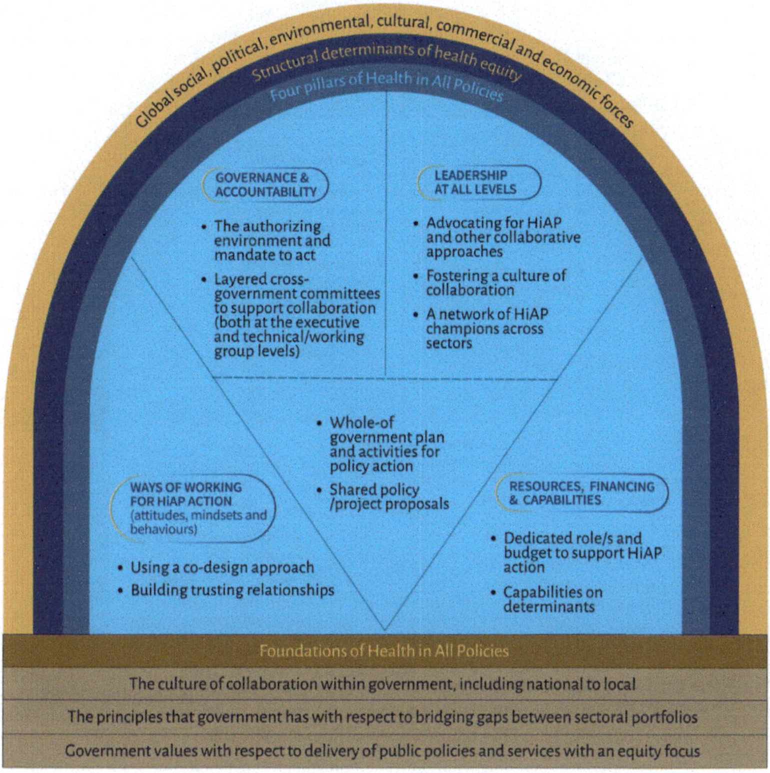

Fig. 1 WHO Health in All Policies 4 Pillars Model

Nonetheless, as per the HiAP experience, there are numerous potential challenges to the application of EiAP, including political will, competing priorities, and the traditional siloed approach to policy problems. The potential to combine the new economic models that account for the real economic value of trees and green space, discussed above, together with the establishment of joined-up government decision processes, in line with the proposed EiAP 4 Pillars model, creates the opportunity to significantly influence government decision-making and public policy.

Despite growing recognition of significance of joined-up public policy approaches to tackle environmental issues such as the provision of green space, multisectoral approaches such as the proposed EiAP 4 Pillars Model remain theoretical. There is a pressing need to move beyond conceptual discussions towards practical, scalable governance changes that enable the systemic integration of environmental concerns across all sectors and levels of government.

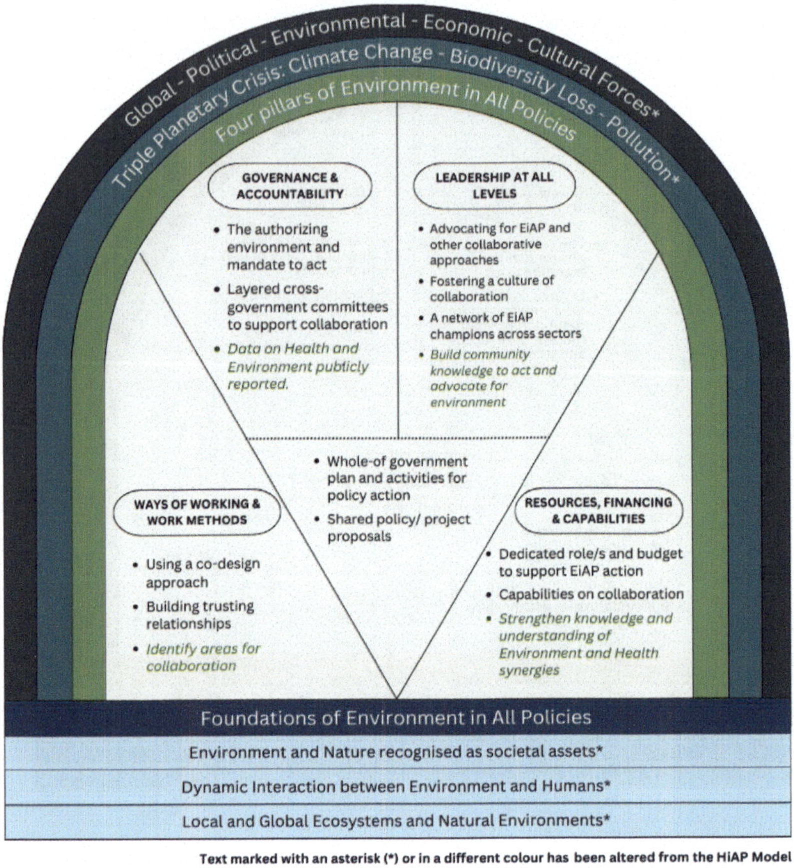

Fig. 2 Environment in All Policies 4 (EiAP) Pillars Model

5.3 *Delivering Equitable Access to Trees and Greenspace*

As this chapter has outlined, insufficient access to the health protecting and promoting benefits of green space is a contributing factor to health inequities. An equity lens that prioritises green space and tree provision in currently underserved areas is vital for reducing inequalities and mitigating the adverse impacts of climate change in cities (United Nations Environment Programme & International Science Council 2024; van Daalen et al. 2024).

Measurable targets (e.g., 30% tree canopy cover or 90% of the population having access to large green space within a 500 m walking distance) need to be included in government policy to help guide equitable delivery of green space. Policy targets should be based on evidence and best-practice and be sufficiently ambitious to drive a step-change in green space access. Targets should be accompanied by specific and coherent actions, legal powers, and budgets at all levels of government—otherwise they remain aspirational.

Green space indicators can inform priority-setting and help monitor policy implementation. Spatial indicators can map the distribution of green space to reveal inequities, monitor changes in green space access over time, and support its retention and expansion amid rapid urban growth and development pressures (Hunter et al. 2023). A range of research- and practice-based metrics have been used to assess the spatial distribution of urban greening and urban green space (Battiston and Schifanella 2024; Browning et al. 2024; UN-Habitat 2022). The World Health Organization (2016) suggests that availability and access are key green space parameters that should be measured and mapped to inform policy. In addition, policy indicators, such as those included in the Global Observatory of Healthy and Sustainable Cities, can monitor policy quality to help fill policy gaps and strengthen targets (Global Healthy and Sustainable City-Indicators Collaboration 2025). Partnerships between policymakers and researchers can ensure that indicators are valid, evidence-informed, and policy-relevant. Levels of government in charge of city planning should regularly report progress on key indicators to aid accountability for delivering on policy targets and to generate community and political support for green space planning.

As we have outlined in this chapter, retention and expansion of green space is challenging in the face of urban development pressures and contestation over how best to allocate land uses. Making room for green space and trees in areas of greatest need, such as lower-income neighbourhoods, requires community and political support. However, communities may oppose green space expansion or request tree removal because of perceived negative impacts on amenity (Roman et al. 2021). Greater understanding of political barriers and the reasons behind community opposition to green space is key to improving equity of access. For example, further research into community knowledge and attitudes to green space and trees, especially in low-income neighbourhoods, could inform approaches to increase awareness of the positive impacts of green space, and support advocacy for change (Fischer et al. 2020; Kelly et al. 2022).

Advocacy efforts led by researchers, community organisations, and governments should highlight the many economic, social, environmental, climate, and health co-benefits of trees and biodiverse green spaces (Hunter et al. 2023; United Nations Environment Programme & International Science Council 2024). The growing interest in city-level climate action plans internationally is another opportunity to embed green space promotion as a key aspect of climate change adaptation and mitigation (Mullenbach and Wilhelm Stanis 2024).

6 Conclusions

Abundant evidence has demonstrated the benefits of green spaces in urban environments, both for nature and human health and wellbeing. Despite this, incorporating quality green space in urban planning and development continues to face challenges that result not only in inequity in the distribution of quality green space, but also the decline of the amount of green space in cities. We have highlighted key issues that trigger these challenges, while acknowledging that there are other factors not discussed in detail in this chapter.

We have shown that green space is often undervalued in urban planning because traditional cost–benefit analysis frameworks in urban management overlook indirect and intangible ecosystem services. Additionally, unavoidable population growth and development pressures in cities have resulted in loss of green spaces through densification and major infrastructure developments, vegetation clearing for new housing, and excessive tree removal.

The challenges must be overcome to implement, maintain, sustain, and improve the distribution and quality of green space. To move forward, we propose three main strategies. First, the economic modelling of green space must be improved to ensure that it is relevant and sensitive to each particular urban context. Improved economic modelling can help policymakers make evidence-informed decisions about green space provision, with a focus on developing high quality green spaces (e.g., spaces that are numerous, large, biodiverse, and provide amenities and ecosystem services). Second, as existing compartmentalised planning systems have not been able to halt or reverse green space losses, there needs to be supportive mechanisms for different parts and levels of government impacting green space to collaborate, operate, and speak the same language. We propose implementing an Environment in All Policies approach, which builds on the strengths of established Health in All Policies approaches, to ensure that both environmental and health outcomes from green space are optimised. Third, evidence-based measurable targets (for example, in terms of tree canopy cover and equitable access to quality green space across different communities) and indicators used to monitor progress need to be included in government policies to help guide equitable delivery of green space. These solutions can only be effectively achieved with genuine collaborations between researchers, community organisations, and governments.

References

ABC News (2022) Adelaide's disappearing tree canopy the subject of new parliamentary inquiry. ABC News. https://www.abc.net.au/news/2022-12-02/adelaide-tree-canopy-subject-of-new-parliamentary-inquiry/101723682

ABC News (2025) New Brisbane stadium to be built at Victoria Park for 2032 Olympic and Paralympic Games. ABC News. https://www.abc.net.au/news/2025-03-25/brisbane-olympics-olympic-games-2032-stadium-premier-plans/105062348

Aerts R, Honnay O, Van Nieuwenhuyse A (2018) Biodiversity and human health: mechanisms and evidence of the positive health effects of diversity in nature and green spaces. Br Med Bull 127(1):5–22. https://doi.org/10.1093/bmb/ldy021

Akaraci S, Feng X, Suesse T, Jalaludin B, Astell-Burt T (2022) Associations between green space, air pollution and birthweight in Sydney Metropolitan Area, Australia. Urban For Urban Green 76:127726. https://doi.org/10.1016/j.ufug.2022.127726

Almanza E, Jerrett M, Dunton G, Seto E, Pentz MA (2012) A study of community design, greenness, and physical activity in children using satellite, GPS and accelerometer data. Health Place 18(1):46–54. https://doi.org/10.1016/j.healthplace.2011.09.003

Andersen L, Corazon SS, Stigsdotter UK (2021) Nature exposure and its effects on immune system functioning: a systematic review. Int J Environ Res Public Health 18(4):1416. https://www.mdpi.com/1660-4601/18/4/1416

Andrews JG, Ammirati RJ, Andrews CJM (2024) Birding benefits: a review of mental health, cognitive, social, and conservation impacts. Ecopsychology 17(1):1–17. https://doi.org/10.1089/eco.2024.0019

Arnberger A (2012) Urban densification and recreational quality of public urban green spaces—a Viennese case study. Sustainability 4(4):703–720. https://www.mdpi.com/2071-1050/4/4/703

Astell-Burt T, Feng X (2019) Association of urban green space with mental health and general health among adults in Australia. JAMA Netw Open 2(7):e198209. https://doi.org/10.1001/jamanetworkopen.2019.8209

Astell-Burt T, Feng X (2020) Urban green space, tree canopy and prevention of cardiometabolic diseases: a multilevel longitudinal study of 46 786 Australians. Int J Epidemiol 49(3):926–933. https://doi.org/10.1093/ije/dyz239

Astell-Burt T, Feng X, Mavoa S, Badland HM, Giles-Corti B (2014) Do low-income neighbourhoods have the least green space? A cross-sectional study of Australia's most populous cities. BMC Public Health 14(1):292. https://doi.org/10.1186/1471-2458-14-292

Astell-Burt T, Hartig T, Eckermann S, Nieuwenhuijsen M, McMunn A, Frumkin H, Feng X (2022) More green, less lonely? A longitudinal cohort study. Int J Epidemiol 51(1):99–110. https://doi.org/10.1093/ije/dyab089

Astell-Burt T, Navakatikyan MA, Feng X (2024) Contact with nature may be a remedy for loneliness: a nationally representative longitudinal cohort study. Environ Res 263:120016. https://doi.org/10.1016/j.envres.2024.120016

Australian Bureau of Statistics (2020) Australians building houses on smaller blocks. Australian Bureau of Statistics. https://www.abs.gov.au/articles/australians-building-houses-smaller-blocks

Australian Urban Observatory (2025) Scorecards. Australian Urban Observatory. https://auo.org.au/measure/scorecards/

Bai X, Nath I, Capon A, Hasan N, Jaron D (2012) Health and wellbeing in the changing urban environment: complex challenges, scientific responses, and the way forward. Curr Opin Environ Sustain 4(4):465–472. https://doi.org/10.1016/j.cosust.2012.09.009

Barton J, Rogerson M (2017) The importance of greenspace for mental health. Bjpsych Int 14(4):79–81. https://doi.org/10.1192/s2056474000002051

Battiston A, Schifanella R (2024) On the need for a multi-dimensional framework to measure accessibility to urban green. NPJ Urban Sustain 4(1):10. https://doi.org/10.1038/s42949-024-00147-y

Bell SL, Michael W, Rebecca L, Wheeler BW (2018) Everyday green space and experienced well-being: the significance of wildlife encounters. Landsc Res 43(1):8–19. https://doi.org/10.1080/01426397.2016.1267721

Bratman GN, Hamilton JP, Hahn KS, Daily GC, Gross JJ (2015) Nature experience reduces rumination and subgenual prefrontal cortex activation. Proc Natl Acad Sci 112(28):8567–8572. https://doi.org/10.1073/pnas.1510459112

Browning MHEM, Locke DH, Konijnendijk C, Labib SM, Rigolon A, Yeager R, Bardhan M, Berland A, Dadvand P, Helbich M, Li F, Li H, James P, Klompmaker J, Reuben A, Roman LA, Tsai WL, Patwary M, O'Neil-Dunne J, Ossola A, Nieuwenhuijsen M (2024) Measuring the 3-30-300 rule to help cities meet nature access thresholds. Sci Total Environ 907:167739. https://doi.org/10.1016/j.scitotenv.2023.167739

Callaghan CT, Major RE, Lyons MB, Martin JM, Kingsford RT (2018) The effects of local and landscape habitat attributes on bird diversity in urban greenspaces. Ecosphere 9(7):e02347. https://doi.org/10.1002/ecs2.2347

Cameron RWF, Brindley P, Mears M, McEwan K, Ferguson F, Sheffield D, Jorgensen A, Riley J, Goodrick J, Ballard L, Richardson M (2020) Where the wild things are! Do urban green spaces with greater avian biodiversity promote more positive emotions in humans? Urban Ecosystems 23(2):301–317. https://doi.org/10.1007/s11252-020-00929-z

Carrus G, Scopelliti M, Lafortezza R, Colangelo G, Ferrini F, Salbitano F, Agrimi M, Portoghesi L, Semenzato P, Sanesi G (2015) Go greener, feel better? The positive effects of biodiversity on the well-being of individuals visiting urban and peri-urban green areas. Land Urban Plan 134:221–228. https://doi.org/10.1016/j.landurbplan.2014.10.022

Chamberlain DE, Gough S, Vaughan H, Vickery JA, Appleton GF (2007) Determinants of bird species richness in public green spaces. Bird Study 54(1):87–97. https://doi.org/10.1080/00063650709461460

Chelli A, Brander L, Geneletti D (2025) Cost-benefit analysis of urban nature-based solutions: a systematic review of approaches and scales with a focus on benefit valuation. Ecosyst Serv 71:101684. https://doi.org/10.1016/j.ecoser.2024.101684

Clark C, Ordóñez C, Livesley SJ (2020) Private tree removal, public loss: valuing and enforcing existing tree protection mechanisms is the key to retaining urban trees on private land. Landsc Urban Plan 203:103899. https://doi.org/10.1016/j.landurbplan.2020.103899

Conway TM (2016) Tending their urban forest: residents' motivations for tree planting and removal. Urban For Urban Green 17:23–32. https://doi.org/10.1016/j.ufug.2016.03.008

Coombes E, Jones AP, Hillsdon M (2010) The relationship of physical activity and overweight to objectively measured green space accessibility and use. Soc Sci Med 70(6):816–822. https://doi.org/10.1016/j.socscimed.2009.11.020

Creed C, Carvalho J (2024) Exploring the user experience, quality, and provision of urban greenspace: a mixed-method approach. Urban For Urban Green 100:1–11. https://doi.org/10.1016/j.ufug.2024.128470

Croeser T, Sharma R, Weisser WW, Bekessy SA (2024) Acute canopy deficits in global cities exposed by the 3–30-300 benchmark for urban nature. Nat Commun 15(1):9333. https://doi.org/10.1038/s41467-024-53402-2

Cui H, Maliki NZ, Wang Y (2024) The role of urban parks in promoting social interaction of older adults in China. Sustainability 16(5):2088. https://www.mdpi.com/2071-1050/16/5/2088

Dai D (2011) Racial/ethnic and socioeconomic disparities in urban green space accessibility: where to intervene? Landsc Urban Plann 102(4):234–244. https://doi.org/10.1016/j.landurbplan.2011.05.002

Dale S (2018) Urban bird community composition influenced by size of urban green spaces, presence of native forest, and urbanization. Urban Ecosyst 21(1):1–14. https://doi.org/10.1007/s11252-017-0706-x

Davern M, Farrar A, Kendal D, Giles-Corti B (2024) Quality green space supporting health, wellbeing and biodiversity: a literature review. https://doi.org/10.25439/rmt.27645495.v1

de Keijzer C, Gascon M, Nieuwenhuijsen MJ, Dadvand P (2016) Long-term green space exposure and cognition across the life course: a systematic review. Curr Environ Health Rep 3(4):468–477. https://doi.org/10.1007/s40572-016-0116-x

de Toledo MCB, Donatelli RJ, Batista GT (2012) Relation between green spaces and bird community structure in an urban area in Southeast Brazil. Urban Ecosyst 15(1):111–131. https://doi.org/10.1007/s11252-011-0195-2

de Vries S, Buijs AE, Snep RPH (2020) Environmental justice in the netherlands: presence and quality of greenspace differ by socioeconomic status of neighbourhoods. Sustainability 12(15):5889. https://www.mdpi.com/2071-1050/12/15/5889

Ekkel ED, de Vries S (2017) Nearby green space and human health: Evaluating accessibility metrics. Landsc Urban Plan 157:214–220. https://doi.org/10.1016/j.landurbplan.2016.06.008

European Environment Agency (2022) Who benefits from nature in cities? Social inequalities in access to urban green and blue spaces across Europe [Briefing]. European Environment Agency. https://www.eea.europa.eu/publications/who-benefits-from-nature-in/who-benefits-from-nature-in

Feng X, Navakatikyan M, Eckermann S, Astell-Burt T (2024) Show me the money! Associations between tree canopy and hospital costs in cities for cardiovascular disease events in a longitudinal cohort study of 110,134 participants. Environ Int 185:108558. https://doi.org/10.1016/j.envint.2024.108558

Ferenc M, Sedláček O, Fuchs R (2014) How to improve urban greenspace for woodland birds: site and local-scale determinants of bird species richness. Urban Ecosyst 17(2):625–640. https://doi.org/10.1007/s11252-013-0328-x

Fischer LK, Neuenkamp L, Lampinen J, Tuomi M, Alday JG, Bucharova A, Cancellieri L, Casado-Arzuaga I, Čeplová N, Cerveró L, Deák B, Eriksson O, Fellowes MDE, Fernández de Manuel B, Filibeck G, González-Guzmán A, Hinojosa MB, Kowarik I, Lumbierres B, Miguel A, Pardo R, Pons X, Rodríguez-García E, Schröder R, Sperandii MG, Unterweger P, Vázquez V, Klaus, VH (2020). Public attitudes toward biodiversity-friendly greenspace management in Europe. Conserv Lett 13(4):e12718. https://doi.org/10.1111/conl.12718

Francis J, Wood LJ, Knuiman M, Giles-Corti B (2012) Quality or quantity? Exploring the relationship between Public Open Space attributes and mental health in Perth, Western Australia. Soc Sci Med 74(10):1570–1577. https://doi.org/10.1016/j.socscimed.2012.01.032

Giles-Corti B, Broomhall MH, Knuiman M, Collins C, Douglas K, Ng K, Lange A, Donovan RJ (2005) Increasing walking: how important is distance to, attractiveness, and size of public open space? Am J Prev Med 28(2, Supplement 2):169–176. https://doi.org/10.1016/j.amepre.2004.10.018

Global Healthy and Sustainable City-Indicators Collaboration (2025) Global observatory of healthy and sustainable cities. Washington University in St. Louis. https://www.healthysustainablecities.org/. Accessed 10 Jan 2025

Green Adelaide (2024) New data shows extent of Adelaide's tree canopy. https://www.greenadelaide.sa.gov.au/news/2024-metro-adelaide-tree-canopy-data

He D, Miao J, Lu Y, Song Y, Chen L, Liu Y (2022) Urban greenery mitigates the negative effect of urban density on older adults' life satisfaction: evidence from Shanghai, China. Cities 124:103607. https://doi.org/10.1016/j.cities.2022.103607

Hoffimann E, Barros H, Ribeiro AI (2017) Socioeconomic inequalities in green space quality and accessibility—evidence from a Southern European city. Int J Environ Res Public Health 14(8):916. https://www.mdpi.com/1660-4601/14/8/916

Holt EW, Lombard QK, Best N, Smiley-Smith S, Quinn JE (2019) Active and passive use of green space, health, and well-being amongst university students. Int J Environ Res Public Health 16(3):424. https://www.mdpi.com/1660-4601/16/3/424

Honey-Rosés J, Zapata O (2023) Green spaces with fewer people improve self-reported affective experience and mood. Int J Environ Res Public Health 20(2):1219. https://www.mdpi.com/1660-4601/20/2/1219

Houlden V, Jani A, Hong A (2021) Is biodiversity of greenspace important for human health and wellbeing? A bibliometric analysis and systematic literature review. Urban For Urban Green 66:127385. https://doi.org/10.1016/j.ufug.2021.127385

Hsu Y-Y, Hawken S, Sepasgozar S, Lin Z-H (2022) Beyond the backyard: GIS analysis of public green space accessibility in Australian metropolitan areas. Sustainability 14(8). https://doi.org/10.3390/su14084694

Hu C-Y, Yang X-J, Gui S-Y, Ding K, Huang K, Fang Y, Jiang Z-X, Zhang X-J (2021) Residential greenness and birth outcomes: a systematic review and meta-analysis of observational studies. Environ Res 193:110599. https://doi.org/10.1016/j.envres.2020.110599

Humphrey JE, Haslem A, Bennett AF (2024) The noisy neighbor conundrum: what influences the value of urban sites for forest birds? Urban Ecosyst 27(3):717–732. https://doi.org/10.1007/s11252-023-01478-x

Hunter RF, Nieuwenhuijsen M, Fabian C, Murphy N, O'Hara K, Rappe E, Sallis JF, Lambert EV, Duenas OLS, Sugiyama T, Kahlmeier S (2023) Advancing urban green and blue space contributions to public health. Lancet Public Health 8(9):e735–e742. https://doi.org/10.1016/S2468-2667(23)00156-1

Ikin K, Le Roux DS, Rayner L, Villaseñor NR, Eyles K, Gibbons P, Manning AD, Lindenmayer DB (2015) Key lessons for achieving biodiversity-sensitive cities and towns. Ecol Manage Restor 16(3):206–214. https://doi.org/10.1111/emr.12180

Irvine KN, Warber SL, Devine-Wright P, Gaston KJ (2013) Understanding urban green space as a health resource: A qualitative comparison of visit motivation and derived effects among park users in Sheffield, UK. Int J Environ Res Public Health 10(1):417–442. https://www.mdpi.com/1660-4601/10/1/417

Islam A, Pattnaik N, Moula MM, Rötzer T, Pauleit S, Rahman MA (2024) Impact of urban green spaces on air quality: a study of PM10 reduction across diverse climates. Sci Total Environ 955:176770. https://doi.org/10.1016/j.scitotenv.2024.176770

Jaganmohan M, Knapp S, Buchmann CM, Schwarz N (2016) The bigger, the better? The influence of urban green space design on cooling effects for residential areas. J Environ Qual 45(1):134–145. https://doi.org/10.2134/jeq2015.01.0062

Jamalishahni T, Turrell G, Foster S, Davern M, Villanueva K (2023) Neighbourhood socio-economic disadvantage and loneliness: the contribution of green space quantity and quality. BMC Public Health 23(1):598. https://doi.org/10.1186/s12889-023-15433-0

Jansen FM, Ettema DF, Kamphuis CBM, Pierik FH, Dijst MJ (2017) How do type and size of natural environments relate to physical activity behavior? Health & Place 46:73–81. https://doi.org/10.1016/j.healthplace.2017.05.005

Jeanes R, Ramón S, Dawn P, O'Connor J (2019) Managing informal sport participation: tensions and opportunities. Int J Sport Policy Polit 11(1):79–95. https://doi.org/10.1080/19406940.2018.1479285

Jimenez MP, DeVille NV, Elliott EG, Schiff JE, Wilt GE, Hart JE, James P (2021) Associations between nature exposure and health: a review of the evidence. Int J Environ Res Public Health 18(9). https://doi.org/10.3390/ijerph18094790

John EE, Astell-Burt T, Yu P, Brennan-Horley C, Feng X (2023) Green space type and healthy ageing in place: an Australian longitudinal study. Urban For Urban Green 84:127903. https://doi.org/10.1016/j.ufug.2023.127903

Kaczynski AT, Potwarka LR, Smale Bryan JA, Havitz ME (2009) Association of Parkland proximity with neighborhood and park-based physical activity: variations by gender and age. Leis Sci 31(2):174–191. https://doi.org/10.1080/01490400802686045

Kandulu J, Soebarto V (2023) Estimating the net benefit of street trees in Greater Adelaide Metropolitan, Australia. https://doi.org/10.25909/24104037.v2

Kelly D, Davern M, Farahani L, Higgs C, Maller C (2022) Urban greening for health and wellbeing in low-income communities: a baseline study in Melbourne, Australia. Cities 120:103442. https://doi.org/10.1016/j.cities.2021.103442

Kirkpatrick JB, Davison A, Daniels GD (2012) Resident attitudes towards trees influence the planting and removal of different types of trees in eastern Australian cities. Landsc Urban Plan 107(2):147–158. https://doi.org/10.1016/j.landurbplan.2012.05.015

Knight T, Price S, Bowler D, Hookway A, King S, Konno K, Richter RL (2021) How effective is 'greening' of urban areas in reducing human exposure to ground-level ozone concentrations, UV exposure and the 'urban heat island effect'? An updated systematic review. Environ Evid 10(1):12. https://doi.org/10.1186/s13750-021-00226-y

Konijnendijk CC (2023) Evidence-based guidelines for greener, healthier, more resilient neighbourhoods: introducing the 3–30–300 rule. J for Res 34(3):821–830. https://doi.org/10.1007/s11676-022-01523-z

Koohsari MJ, Akitomo Y, Jenny V, Kaczynski AT, Oka K (2024) The density paradox: density, walking, and psychological stress in overcrowded public spaces *Cities & Health* 1–4. https://doi.org/10.1080/23748834.2024.2391222

Kronenberg J, Haase A, Łaszkiewicz E, Antal A, Baravikova A, Biernacka M, Dushkova D, Filčak R, Haase D, Ignatieva M, Khmara Y, Niţă MR, Onose DA (2020) Environmental justice in the context of urban green space availability, accessibility, and attractiveness in postsocialist cities. Cities 106:102862. https://doi.org/10.1016/j.cities.2020.102862

Kuddus MA, Tynan E, McBryde E (2020) Urbanization: a problem for the rich and the poor? Public Health Rev 41(1):1. https://doi.org/10.1186/s40985-019-0116-0
Lambert A, Vlaar J, Herrington S, Brussoni M (2019) What is the relationship between the neighbourhood built environment and time spent in outdoor play? A systematic review. Int J Environ Res Public Health 16(20):3840. https://www.mdpi.com/1660-4601/16/20/3840
Lepczyk CA, Aronson MFJ, Evans KL, Goddard MA, Lerman SB, MacIvor JS (2017) Biodiversity in the city: fundamental questions for understanding the ecology of urban green spaces for biodiversity conservation. Bioscience 67(9):799–807. https://doi.org/10.1093/biosci/bix079
Liu D, Kwan M-P, Kan Z (2021) Analysis of urban green space accessibility and distribution inequity in the city of Chicago. Urban For Urban Green 59:127029. https://doi.org/10.1016/j.ufug.2021.127029
Local Government Association of South Australia (2024) New tree protections to increase Adelaide's tree canopy. https://www.lga.sa.gov.au/news-and-events/news/latest-news/2024/new-rules-to-protect-adelaide-tree-canopy
Lu P, van Ameijde J (2025) The impact of high-density urban environments on children's play, a systematic review of current insights. Cities & Health 1–20. https://doi.org/10.1080/23748834.2024.2441554
Luo YN, Huang WZ, Liu XX, Markevych I, Bloom MS, Zhao T, Heinrich J, Yang BY, Dong GH (2020) Greenspace with overweight and obesity: A systematic review and meta-analysis of epidemiological studies up to 2020 [Review]. Obes Rev 21(11):Article e13078. https://doi.org/10.1111/obr.13078
Maas J, Verheij RA, Spreeuwenberg P, Groenewegen PP (2008) Physical activity as a possible mechanism behind the relationship between green space and health: a multilevel analysis. BMC Public Health 8(1):206. https://doi.org/10.1186/1471-2458-8-206
Maller C, Townsend M, Pryor A, Brown P, St Leger L (2006) Healthy nature healthy people: 'contact with nature' as an upstream health promotion intervention for populations. Health Promot Int 21(1):45–54. https://doi.org/10.1093/heapro/dai032
Masini E, Tomao A, Corona P, Fattorini L, Giuliarelli D, Portoghesi L, Agrimi M (2023) The ecosystem disservices of trees on sidewalks: a study based on a municipality urban tree inventory in Central Italy. Urban For Urban Green 86:128007. https://doi.org/10.1016/j.ufug.2023.128007
Mata L, Andersen AN, Morán-Ordóñez A, Hahs AK, Backstrom A, Ives CD, Bickel D, Duncan D, Palma E, Thomas F, Cranney K, Walker K, Shears I, Semeraro L, Malipatil M, Moir ML, Plein M, Porch N, Vesk PA, Smith TR, Lynch Y (2021) Indigenous plants promote insect biodiversity in urban greenspaces. Ecol Appl 31(4):e02309. https://doi.org/10.1002/eap.2309
Mitchell RJ, Richardson EA, Shortt NK, Pearce JR (2015) Neighborhood environments and socioeconomic inequalities in mental well-being. Am J Prev Med 49(1):80–84. https://doi.org/10.1016/j.amepre.2015.01.017
Mullenbach LE, Wilhelm Stanis SA (2024) Climate change adaptation plans: inclusion of health, equity, and green space. J Urban Aff 46(4):701–716. https://doi.org/10.1080/07352166.2022.2091449
Mushangwe S, Astell-Burt T, Steel D, Feng X (2021) Ethnic inequalities in green space availability: evidence from Australia. Urban for Urban Green 64:127235. https://doi.org/10.1016/j.ufug.2021.127235
Mygind L, Kjeldsted E, Hartmeyer R, Mygind E, Stevenson MP, Quintana DS, Bentsen P (2021) Effects of public green space on acute psychophysiological stress response: a systematic review and meta-analysis of the experimental and quasi-experimental evidence. Environ Behav 53(2):184–226. https://doi.org/10.1177/0013916519873376
Mytton OT, Townsend N, Rutter H, Foster C (2012) Green space and physical activity: an observational study using Health Survey for England data. Health Place 18(5):1034–1041. https://doi.org/10.1016/j.healthplace.2012.06.003
North East Link (2025) Greener north east. https://bigbuild.vic.gov.au/projects/north-east-link/about/environment/tree-planting-program
Nowak DJ (2021). Understanding i-Tree. https://doi.org/10.2737/NRS-GTR-200-2021

NSW Department of Planning (2025) Cumberland Plain Conservation Plan. https://www.planning.nsw.gov.au/policy-and-legislation/strategic-conservation-planning/cumberland-plain-conservation-plan
Ordóñez-Barona C, Bush J, Hurley J, Amati M, Juhola S, Frank S, Ritchie M, Clark C, English A, Hertzog K, Caffin M, Watt S, Livesley SJ (2021) International approaches to protecting and retaining trees on private urban land. J Environ Manage 285:112081. https://doi.org/10.1016/j.jenvman.2021.112081
Otto J, Borgström S, Haase D, Andersson E (2024) Capturing residents' perceptions of green spaces in densifying urban landscapes—the potentials of mental mapping. Urban For Urban Green 94:128266. https://doi.org/10.1016/j.ufug.2024.128266
Parsons CR, Cunningham S, Le Roux DS, Gibbons P (2023) The impact of greenfield residential development on mature trees. Urban For Urban Green 86:128006. https://doi.org/10.1016/j.ufug.2023.128006
Peterson MN, Larson LR, Hipp A, Beall JM, Lerose C, Desrochers H, Lauder S, Torres S, Tarr NA, Stukes K, Stevenson K, Martin KL (2024) Birdwatching linked to increased psychological well-being on college campuses: a pilot-scale experimental study. J Environ Psychol 96:102306. https://doi.org/10.1016/j.jenvp.2024.102306
Pihkala P (2020) Anxiety and the ecological crisis: an analysis of eco-anxiety and climate anxiety. Sustainability 12(19):7836. https://www.mdpi.com/2071-1050/12/19/7836
Ramirez-Rubio O, Daher C, Fanjul G, Gascon M, Mueller N, Pajín L, Plasencia A, Rojas-Rueda D, Thondoo M, Nieuwenhuijsen MJ (2019) Urban health: an example of a "health in all policies" approach in the context of SDGs implementation. Glob Health 15(1):87. https://doi.org/10.1186/s12992-019-0529-z
Rigolon A, Browning MHEM, McAnirlin O, Yoon H (2021) Green space and health equity: a systematic review on the potential of green space to reduce health disparities. Int J Environ Res Public Health 18(5):2563. https://www.mdpi.com/1660-4601/18/5/2563
Rojas-Rueda D, Nieuwenhuijsen MJ, Gascon M, Perez-Leon D, Mudu P (2019) Green spaces and mortality: a systematic review and meta-analysis of cohort studies. Lancet Planetary Health 3(11):e469–e477. https://doi.org/10.1016/S2542-5196(19)30215-3
Roman LA, Conway TM, Eisenman TS, Koeser AK, Ordóñez Barona C, Locke DH, Jenerette GD, Östberg J, Vogt J (2021) Beyond 'trees are good': Disservices, management costs, and tradeoffs in urban forestry. Ambio 50(3):615–630. https://doi.org/10.1007/s13280-020-01396-8
Rosario LD, Astell-Burt T, Navakatikyan M, Olsen JR, Caryl F, Lin B, Jalaludin B, de Leeuw E, Mitchell R, Feng X (2025) Socioeconomic inequities within and between cities in objectively measured green space qualities at small geographical scales: Evidence from Australia. medRxiv. https://doi.org/10.1101/2025.04.09.25325554
Russell K (2021) Potential sediment supply fluxes associated with greenfield residential construction. Anthropocene 35:100300. https://doi.org/10.1016/j.ancene.2021.100300
Schüle SA, Gabriel KMA, Bolte G (2017) Relationship between neighbourhood socioeconomic position and neighbourhood public green space availability: an environmental inequality analysis in a large German city applying generalized linear models. Int J Hyg Environ Health 220(4):711–718. https://doi.org/10.1016/j.ijheh.2017.02.006
Shanahan DF, Bush R, Gaston KJ, Lin BB, Dean J, Barber E, Fuller RA (2016) Health benefits from nature experiences depend on dose. Sci Rep 6(1):28551. https://doi.org/10.1038/srep28551
Shin JC, Parab KV, An R, Grigsby-Toussaint DS (2020) Greenspace exposure and sleep: a systematic review. Environ Res 182:109081. https://doi.org/10.1016/j.envres.2019.109081
Smith IA, Dearborn VK, Hutyra LR (2019) Live fast, die young: accelerated growth, mortality, and turnover in street trees. PLoS ONE 14(5):e0215846. https://doi.org/10.1371/journal.pone.0215846
Taylor L, Hochuli DF (2017) Defining greenspace: multiple uses across multiple disciplines. Landsc Urban Plan 158:25–38. https://doi.org/10.1016/j.landurbplan.2016.09.024
Tefera Y, Soebarto V, Bishop C, Kandulu J, Williams C (2023) A scoping review of urban planning decision support tools and processes that account for the health, environment, and economic

benefits of trees and greenspace. Int J Environ Res Public Health 21(1). https://doi.org/10.3390/ijerph21010048
Teotónio I, Silva CM, Cruz CO (2021) Economics of green roofs and green walls: a literature review. Sustain Cities Soc 69:102781. https://doi.org/10.1016/j.scs.2021.102781
The World Bank (2023) Urban Development. https://www.worldbank.org/en/topic/urbandevelopment/overview
Threlfall CG, Mata L, Mackie JA, Hahs AK, Stork NE, Williams NSG, Livesley SJ (2017) Increasing biodiversity in urban green spaces through simple vegetation interventions. J Appl Ecol 54(6):1874–1883. https://doi.org/10.1111/1365-2664.12876
Toftager M, Ekholm O, Schipperijn J, Stigsdotter U, Bentsen P, Grønbæk M, Randrup TB, Kamper-Jørgensen F (2011) Distance to Green space and physical activity: a Danish national representative survey. J Phys Act Health 8(6):741–749. https://doi.org/10.1123/jpah.8.6.741
UN-Habitat (2022) The global urban monitoring framework: a guide for urban monitoring of SDGs and NUA and other urban-related thematic or local, national and global frameworks. https://unhabitat.org/the-global-urban-monitoring-framework
United Nations Environment Programme & International Science Council (2024) Navigating new horizons: A global foresight report on planetary health and human wellbeing. https://wedocs.unep.org/20.500.11822/45890
van Daalen KR, Tonne C, Semenza JC, Rocklöv J, Markandya A, Dasandi N, Jankin S, Achebak H, Ballester J, Bechara H, Beck TM, Callaghan MW, Carvalho BM, Chambers J, Pradas MC, Courtenay O, Dasgupta S, Eckelman MJ, Farooq Z, Lowe R et al (2024) The 2024 Europe report of the Lancet Countdown on health and climate change: unprecedented warming demands unprecedented action. Lancet Public Health (published online May 12). https://doi.org/10.1016/S2468-2667(24)00055-0
van Dillen SME, de Vries S, Groenewegen PP, Spreeuwenberg P (2012) Greenspace in urban neighbourhoods and residents' health: adding quality to quantity. J Epidemiol Community Health 66(6):e8. https://doi.org/10.1136/jech.2009.104695
Van Renterghem, T. (2024). In-situ co-creative soundscape augmentation with fountains: balancing sound preference and road traffic noise masking in urban parks. Appl Acoust 222:110001. https://doi.org/10.1016/j.apacoust.2024.110001
Victorian Auditor-General's Office (2020) Protecting critically endangered Grasslands. https://www.audit.vic.gov.au/report/protecting-critically-endangered-grasslands?section=
Viinikka A, Tiitu M, Heikinheimo V, Halonen JI, Nyberg E, Vierikko K (2023) Associations of neighborhood-level socioeconomic status, accessibility, and quality of green spaces in Finnish urban regions. Appl Geogr 157:102973. https://doi.org/10.1016/j.apgeog.2023.102973
White MP, Alcock I, Grellier J, Wheeler BW, Hartig T, Warber SL, Bone A, Depledge MH, Fleming LE (2019) Spending at least 120 minutes a week in nature is associated with good health and wellbeing. Sci Rep 9(1):7730. https://doi.org/10.1038/s41598-019-44097-3
Williams C, Smith JA, Valentine N, Baum F, Friel S, Williams J, Schmitt D (2023) The wellbeing economy and health in all policies: fostering action for change. Health Promot J Austr 34(3):623–625. https://doi.org/10.1002/hpja.768
Wolch JR, Byrne J, Newell JP (2014) Urban green space, public health, and environmental justice: the challenge of making cities 'just green enough'. Landsc Urban Plann 125:234–244. https://doi.org/10.1016/j.landurbplan.2014.01.017
Wolf KL, Lam ST, McKeen JK, Richardson GRA, van den Bosch M, Bardekjian AC (2020) Urban trees and human health: a scoping review. Int J Environ Res Public Health 17(12). https://doi.org/10.3390/ijerph17124371
Wood E, Harsant A, Dallimer M, Cronin de Chavez A, McEachan RRC, Hassall C (2018) Not all green space is created equal: biodiversity predicts psychological restorative benefits from urban green space [Original Research]. Front Psychol 9. https://doi.org/10.3389/fpsyg.2018.02320
World Health Organization (2014) Health in all policies: Helsinki statement. Framework for country action. https://www.who.int/publications/i/item/9789241506908

World Health Organization (2016) Urban green spaces and health. World Health Organization. Regional Office for Europe. https://iris.who.int/handle/10665/345751

World Health Organization (2017) Urban green spaces: a brief for action. World Health Organization. https://www.who.int/europe/publications/i/item/9789289052498

World Health Organization (2023) Working together for equity and healthier populations: sustainable multisectoral collaboration based on Health in All Policies approaches. https://www.who.int/publications/i/item/9789240067530

Wright Wendel HE, Zarger RK, Mihelcic JR (2012) Accessibility and usability: green space preferences, perceptions, and barriers in a rapidly urbanizing city in Latin America. Landsc Urban Plan 107(3):272–282. https://doi.org/10.1016/j.landurbplan.2012.06.003

Yang B-Y, Zhao T, Hu L-X, Browning MHEM, Heinrich J, Dharmage SC, Jalaludin B, Knibbs LD, Liu X-X, Luo Y-N, James P, Li S, Huang W-Z, Chen G, Zeng X-W, Hu L-W, Yu Y, Dong G-H (2021) Greenspace and human health: an umbrella review. Innovation 2(4). https://doi.org/10.1016/j.xinn.2021.100164

Yang L, Zhang L, Li Y, Wu S (2015) Water-related ecosystem services provided by urban green space: a case study in Yixing City (China). Landsc. Urban Plann 136:40–51. https://doi.org/10.1016/j.landurbplan.2014.11.016

Yuan Y, Huang F, Lin F, Zhu P, Zhu P (2021) Green space exposure on mortality and cardiovascular outcomes in older adults: a systematic review and meta-analysis of observational studies. Aging Clin Exp Res 33(7):1783–1797. https://doi.org/10.1007/s40520-020-01710-0

Zhang Y-D, Zhou G-L, Wang L, Browning MHEM, Markevych I, Heinrich J, Knibbs LD, Zhao T, Ding Y, Chen S, Liu K-K, Dadvand P, Dong G-H, Yang B-Y (2024) Greenspace and human microbiota: a systematic review. Environ Int 187:108662. https://doi.org/10.1016/j.envint.2024.108662

Zhang Y, Mavoa S, Zhao J, Raphael D, Smith M (2020) The association between green space and adolescents' mental well-being: a systematic review. Int J Environ Res Public Health 17(18):6640. https://www.mdpi.com/1660-4601/17/18/6640

Zhang Y, Van den Berg AE, Van Dijk T, Weitkamp G (2017) Quality over quantity: contribution of urban green space to neighborhood satisfaction. Int J Environ Res Public Health 14(5):535. https://www.mdpi.com/1660-4601/14/5/535

Zhang Z, Měchurová K, Resch B, Amegbor P, Sabel CE (2023) Assessing the association between overcrowding and human physiological stress response in different urban contexts: a case study in Salzburg, Austria. Int J Health Geogr 22(1):15. https://doi.org/10.1186/s12942-023-00334-7

Zimmermann E, Bracalenti L, Piacentini R, Inostroza L (2016) Urban flood risk reduction by increasing green areas for adaptation to climate change. Procedia Eng 161:2241–2246. https://doi.org/10.1016/j.proeng.2016.08.822

Fostering Nature Connection and Well-Being Through Nature-Based Planning and Design in a High-Density Tropical City

Angelia Sia, Rachel R. Y. Oh, Jason Wright, Puay Yok Tan, and Kenneth B. H. Er

Abstract It is expected that 70% of the global population will reside in urban areas by 2050, up from the current 56%. Competing demands for land to support economic activities and infrastructure development will reduce the availability of urban natural habitats and recreational green areas. This could contribute to phenomena such as "Nature Deficit Disorder" and the "Extinction of Experience"—terms that describe a growing disconnection from nature, which can have adverse consequences for both environmental health and human well-being. Tropical cities are particularly vulnerable, as urban expansion in these regions often coincides with the rapid loss of native, biodiversity rich habitats. Despite these challenges, we argue that disconnection from nature can be averted through intentional land use planning and strategic nature conservation approaches to sustain and even enhance human-nature interactions, with Singapore providing a compelling case in point. Situated 1° N of the equator, it is one of the world's most densely populated tropical cities yet remains remarkably green. Residents continue to maintain a strong connection with nature, opposing global trends of disconnection. Singapore's success stems from its commitment to preserving and enhancing an interconnected network of green

---

A. Sia (✉) · J. Wright · P. Y. Tan · K. B. H. Er
National Parks Board, Singapore, Singapore
e-mail: angelia_sia@nparks.gov.sg

J. Wright
e-mail: jason_wright@nparks.gov.sg

P. Y. Tan
e-mail: tan_puay_yok@nparks.gov.sg

R. R. Y. Oh
Helmholtz Centre for Environmental Research, UFZ, German Centre for Integrative Biodiversity Research (iDiv) Halle-Jena-Leipzig), Leipzig, Germany
e-mail: r.oh@uq.edu.au

R. R. Y. Oh · P. Y. Tan
National University of Singapore, Singapore, Singapore

A. Cheshmehzangi et al. (eds.), *Nature in Cities, Nurturing Cities*, Urban Sustainability,
https://doi.org/10.1007/978-981-95-5487-4_4

spaces, including nature reserves, gardens, parks, and greenways. These are complemented by the greening of the urban matrix, targeted habitat restoration and biodiversity enhancement projects. Community-led horticultural programmes further deepen residents' engagement with nature, while innovative therapeutic landscape designs harness the restorative benefits of greenery to support human well-being. Drawing on a decade of evidence from Singapore, we evaluate the outcomes of these initiatives and discuss how similar strategies can be adapted for other tropical cities experiencing urbanization pressures.

Keywords Tropical cities · Urban nature · Extinction of experience · Connection with nature · Nature-based interventions · Well-being

1 Introduction: Urbanization and Its Challenges in Tropical Cities

Global urban populations are projected to rise from 56 to 70% by 2050, marking a considerable shift in human settlement patterns (United Nations 2018). This trend poses multifaceted challenges for city planners and policymakers, particularly in tropical regions where urban growth outpaces global averages (State of the Tropics 2020). Currently, approximately 40% of the global population resides in the tropics, and this proportion is expected to grow to 50% by 2050 (ibid.). As urbanization intensifies, many cities and peri-urban areas in the tropics will experience habitat loss, resulting in the loss of unique ecosystems and species (McDonald et al. 2018). These environmental impacts extend beyond local boundaries, affecting the region and the world as a whole (Roth 2007; Tan 2019), and underscore the urgency of integrating sustainability into urban development.

Tropical cities face compounded pressures arising from their geographical context, including rising temperatures that could diminish urban liveability and constrain outdoor activities (Revi et al. 2014). Such conditions demand innovative strategies to harmonize human needs with environmental preservation. Integrating nature into urban planning emerges as an effective strategy to address both environmental sustainability and residents' well-being.

2 The Role of Nature in Human Well-Being

Nature enhances human well-being through multiple pathways, including increasing physical activity, promoting relaxation, reducing stress, and enhancing social interactions (White et al. 2023). However, most of this evidence originates from studies conducted in temperate, high-income regions especially in Europe and North America, while research from the Global South remains comparatively scarce, creating a significant geographic and socio-economic imbalance in the knowledge base (Marvier et al. 2023). Greener neighborhoods have been reported to be associated with a reduced probability of major depressive disorders and hence population well-being (Sankar et al. 2018). Exposure to natural environments further fosters relaxation and lowers stress levels by reducing rumination, thereby supporting psychological resilience (Bratman et al. 2015). Urban green spaces strengthen community bonds through social interactions, thereby contributing to collective emotional well-being (Cardinali et al. 2024). Importantly, these effects have long-term implications—childhood exposure to green spaces predicts a lower risk of psychiatric disorders in adolescence and adulthood (Engemann et al. 2019). These findings highlight the importance of prioritizing nature-integrated urban planning, as a public health and sustainability imperative.

Beyond mere vegetation cover (greenness) and the availability of green spaces, cultivating a strong sense of connection to nature—often termed "nature connectedness" (Mayer and Franz 2004) or "nature relatedness" (Nisbet et al. 2008)—is critical for well-being. The construct, as defined by Nisbet and colleagues, is multifaceted, capturing an individual's cognitive, affective, and experiential connection with the natural world. It reflects not just an affection for nature but also an awareness, understanding, and sense of interconnectedness with all aspects of nature, including those that may be ecologically essential yet aesthetically unappealing or lacking direct benefits to humans. Relatedly, urban expansion and the corresponding loss of natural areas have raised concerns about a growing disconnect between urbanites and nature (Beatley 2011). Modern indoor, sedentary, and technology-driven lifestyles may have further reduced both the frequency and quality of human interactions with nature. Consequently, phenomena such as "Nature Deficit Disorder" (Louv 2005) and the "Extinction of Experience" (Pyle 1978; Soga and Gaston 2016) have emerged—both linked to negative impacts on human health and well-being. However, these effects tend to vary across cultural and geographic contexts (Soga and Gaston 2023).

Significant knowledge gaps persist in our understanding of how different types of human-nature interactions influence outcomes, and the contextual factors that shape them, especially within highly urbanized tropical environments (Oh et al. 2020). Empirical research on the potential benefits of nature exposure in these settings, such as improvements in physical health, emotional well-being, and social connections through engagement with urban green spaces remains underrepresented compared to those conducted in temperate contexts (Oh et al. 2021a, b). These regions face a multitude of interlinked challenges: climate change intensifies urban heat island effects and the frequency and intensity of extreme weather events such as cyclones (Deng et al. 2025). Relatedly, rapid urbanization reduces green space availability, while dense populations place escalating pressures on already limited existing resources. Simultaneously, tropical cities harbor exceptional levels of biodiversity, yet urbanization threatens endemic species and ecosystems (Bonebrake et al. 2024). These interconnected challenges underscore the need for targeted research and strategies to safeguard and expand urban green spaces in tropical contexts.

In this chapter, we examine Singapore's urban greening initiatives, showcasing its innovative approaches to integrate nature into dense urban environments. We explore the broader implications for tropical cities seeking to balance urban growth, greening and human well-being.

3 Singapore as a Case Study in Urban Greening and Equity

Situated just 1^0 N of the equator, Singapore experiences a tropical climate marked by year-round warmth and high rainfall. Despite its small size as a city-state spanning 736 km^2 and ranking among the most densely populated countries globally, Singapore boasts notable biodiversity, with over 3,000 native species across 10 major taxonomic groups (Chisholm et al. 2023). This richness, concentrated within such a compact area, positions Singapore as a biodiversity hotspot relative to its size.

In the early 1800s, Singapore was densely forested. However, the arrival of the British in 1819 marked the beginning of extensive deforestation, primarily driven by timber extraction and the rapid expansion of agricultural activities such as gambier and pepper plantations. By the late nineteenth century, most of the island's primary forest had been cleared, leaving little of the original rainforest intact. Recognizing the severe environmental impacts of this deforestation, the British administration, guided by recommendations from Nathaniel Cantley, established the first Forest Reserves in 1884 to protect the remaining forested areas and promote reforestation efforts (National Parks Board 2015). At this point, less than 7% of the primary forest remained.

Since gaining independence in 1965, Singapore has prioritized urban greening alongside rapid development. While much of the original forest lost to agriculture and urbanization has been replaced by managed vegetation, the city-state today boasts a diverse mosaic of green spaces—including parks, green corridors such as the Park Connector Network, lush streetscape greenery, and skyrise gardens integrated across the urban landscape (Fig. 1).

Singapore presents a unique case study within the global discourse on urban greening, human-nature interactions and environmental equity in tropical environments. Unlike many major cities where access to urban nature is disproportionately concentrated in affluent neighborhoods—often due to property value premiums associated with green spaces (Belcher and Chisholm 2018)—Singapore has achieved a more equitable distribution of urban greenery across socio-economic groups (Nghiem et al. 2021a, b). This outcome is largely attributed to two key policy and structural mechanisms. First, Singapore's housing landscape is shaped by strong state intervention, with approximately 80% of the population residing in public housing governed by explicit quotas that ensure a balanced representation of income levels and ethnicities (Department of Statistics Singapore 2019). Second, Singapore's urban planning framework embeds greenery as a fundamental component of residential design. National targets stipulate that by 2030, all households should be located within a 10-min walk of an urban greenspace—an increase of 10% in accessibility from 2015 (Sustainable Singapore Blueprint 2015).

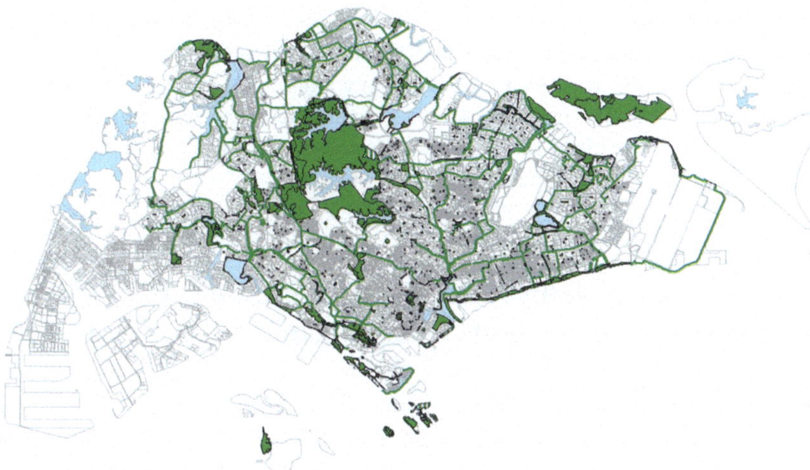

Fig. 1 Singapore's Island-wide Recreational Network (*Source* Urban Redevelopment Authority).

Singapore's current "City in Nature" initiative builds on earlier greening efforts, to further integrate ecology into urban planning to enhance liveability, sustainability, and human well-being (Er 2021). This vision is being realized through three key strategies:

- **Enhancing Nature Connection by Extending Natural Capital**: Efforts include expanding its Nature Reserves (legally protected areas for conserving native biodiversity) by integrating them with Nature Parks (buffer zones surrounding the reserves) and Nature Ways (roadside ecological corridors connecting green spaces). Together, these create a network that interfaces seamlessly with the existing urban greenery. This approach enhances ecological connectivity, safeguards core biodiversity areas, and provides residents with greater opportunities for recreation and connection with nature.
- **Enhancing and Expanding Nature in Urban Gardens and Parks, and Improving Access**: Enhancements encompass habitat restoration, naturalistic landscaping, and the naturalization of waterways, alongside specially designed spaces such as Therapeutic Gardens, Nature Playgardens, and Contemplative Landscapes (see Sect. 4). These features not only promote native biodiversity but also offer diverse settings for relaxation, play, and reflection, contributing to improved well-being.
- **Facilitating Nature Interactions Through Landscape Design and Horticultural Programs**: Beyond infrastructure, targeted programmes encourage residents to engage with green spaces. Initiatives such as community and allotment gardening provide opportunities for engaging with nature, fostering a deeper human-nature connection in a dense urban environment.

Moreover, Singapore's equatorial location ensures year-round vegetation cover, in contrast to the pronounced seasonal fluctuations in temperate climates. This continuous presence of greenery may yield more stable and sustained provision of ecosystem service benefits, particularly for human health and well-being. As such, knowledge from tropical cities like Singapore is critical to advance the evidence base on the long-term impacts of urban greening strategies.

4 Contextualizing Nature and Well-Being Connections in Tropical Cities: Insights from Singapore

Research on the relationship between people and nature in Singapore has been instrumental in developing evidence-based insights into how urban nature fosters nature connectedness (a concept defined earlier in Sect. 2). These studies have also shown that urban nature enhances well-being, which is understood as a state of positive feelings and realizing one's full potential (Simons et al. 2021). The findings have informed Singapore's urban greening strategies, particularly under the "City in Nature" vision (see Sect. 3).

A key premise of the approach is that an individual's sense of connection with nature is malleable and influenced by their physical environment. For instance, studies have shown that rural residents have higher levels of nature connectedness than their urban counterparts (Bashan et al. 2021). This highlights the need to better understand whether Singapore's urban greening efforts and recent nature restoration initiatives have effectively strengthened residents' connection to nature and improved their well-being, as well as how these outcomes compare to those in other cities and regions. Importantly, identifying the specific pathways and interventions within the built environment that facilitate greater nature connectedness, and their positive effects on well-being, is a key focus for ongoing research and urban planning.

4.1 Key Factors of Nature Connectedness in Singapore

The impact of Singapore's public green infrastructure is significantly enhanced as planning and design intentionally address the frequency, duration, and quality of nature exposure—factors identified by Shanahan et al. (2015) as critical for optimizing the health benefits of urban nature. The key strategies of the City in Nature initiative have collectively contributed to strengthening residents' connection with nature, as outlined below.

- **Enhancing Residents' Nature Connection by Extending Natural Capital**

  The positive impact of nature connection on well-being is well-established across numerous studies, although the strength and nature of this relationship can vary according to geographical, cultural, and socio-economic contexts (Soga and

Gaston 2023). While much of the existing evidence comes from temperate cities, recent research in tropical cities—particularly Singapore—has begun to highlight similar patterns.

In Singapore, Oh et al. (2021a, b) found that the relationship between "nature dose" (frequency, duration, and quality of nature exposure) and mental well-being (depression, stress, and anxiety), as well as physical health (high blood pressure and diabetes), is moderated by an individual's connection to nature. Notably, those with a stronger sense of nature connection were less likely to experience depression, stress, and anxiety, regardless of the amount of time spent in parks. Similarly, Sia et al. (2023a, b, 2025) reported that nature connection had a slightly stronger effect on well-being than time spent in parks alone.

The COVID-19 pandemic further underscored the importance of access to nature. During the pandemic period, park usage patterns in Singapore shifted significantly: a survey found a 71% increase in visitors to nature parks, compared to a 21% increase in urban parks, suggesting a marked preference for more naturalistic environments (Yap et al. 2022). This aligns with global findings that pandemic-related stress has heightened people's desire for nature contact (Berdejo-Espinola et al. 2021), and that biodiverse environments are generally associated with improved psychological well-being (Fuller et al. 2007).

However, the relationship between biodiversity and well-being remains complex and, at times, inconclusive. Nghiem et al. (2021a, b) found that perceived animal diversity in nature parks positively influenced park users' connection to nature and mental well-being. In contrast, Sia et al. (2025) observed no such effect from actual bird diversity; in fact, their findings suggest a weak negative direct effect, possibly reflecting ecosystem disservices linked to urban wildlife. Nevertheless, the overall impact on well-being remained positive when mediated by individuals' sense of nature connectedness.

These contrasting findings illustrate the need for further research to disentangle the emotional, experiential, and cognitive drivers of the relationship between human-nature connection and well-being, especially in tropical cities like Singapore.

Key Strategies

Singapore's City in Nature initiative was launched to integrate and enhance natural features within its extensive network of parks and green corridors, bringing people closer to nature. Key efforts include establishing about 450 hectares of nature parks around core reserves and restoring habitats within urban parks. A notable example is the Rifle Range Nature Park (Fig. 2), a 66-ha lowland forest that buffers the Bukit Timah Nature Reserve. The park features over 7 km of boardwalks and trails, many following original *kampung* (traditional village) paths from the 1950s. Designed for inclusivity, it offers child-friendly trails, barrier-free routes for people with disabilities, and a distinctive colugo-shaped viewing deck rising 31 m above a converted quarry wetland, offering panoramic views of the forest canopy.

Fig. 2 Rifle Range Nature Park features over 7 km of boardwalks and trails that cater to people of all ages and abilities

To further strengthen ecological connectivity, Singapore has developed a network of park connectors—recreational pathways linking parks and nature ways, which serve as biodiversity corridors. Both are planted with native trees and shrubs in layered tiers to mimic tropical forest structure (Er 2021; Er and Leong 2023). These initiatives are complemented by an expanding network of nature trails and parks that foster nature connectedness among residents. For instance, the 24 km Rail Corridor (Fig. 3) links several key biodiversity sites, including Rifle Range Nature Park, facilitating wildlife movement and providing a continuous green corridor for recreation. Thoughtfully designed to preserve natural and cultural heritage, the Rail Corridor also features new community nodes, offering diverse opportunities for people to experience and connect with nature.

Fig. 3 The 24 km Rail Corridor connects several core biodiversity sites across Singapore

- **Enhancing and Expanding Nature in Urban Gardens and Parks, and Improving Access**

  Arising from Singapore's extensive park network and well-connected park connector system, a multi-ethnic study of over 3,000 participants revealed high park usage among residents—76% reported monthly park visits with an average park use time of 10.6 hours (Petrunoff et al. 2021). Greater time spent in parks was consistently linked to better well-being, demonstrating a clear dose-response relationship (ibid.). Furthermore, proximity to parks, and hence nature, has been shown to enhance life satisfaction (Chang et al. 2020).

  Comparing between Singapore and Brisbane, Nghiem et al. (2021a, b) found that Singapore has twice as many publicly accessible parks within 500 m of residences. Oh et al. (2020) also compared four types of nature interactions—indirect (window views), incidental, intentional time in private gardens, and intentional time in public green spaces—between the two cities. When total time spent across all four types of green spaces was considered, residents of low-density Brisbane—where private gardens and balconies are more common—spent about twice as much time in nature overall as those in Singapore. However, Singapore residents spent more time specifically in public parks and gardens than their counterparts in Brisbane, likely reflecting the city's intentional planning and high accessibility of public green spaces. In contrast, indirect interactions—such as views of greenery—were less frequent in Singapore as high-rise living often limits such views, underscoring the role of urban design in shaping nature engagement. These findings suggest that differences in residential typologies and private access to natural spaces can notably influence nature engagement patterns.

In a separate study, Sia et al. (2023a, b) found that in Singapore, nature orientation, social norms, and perceived accessibility all influence park visits, with nature orientation being the strongest factor. However, only nature orientation and time spent in parks were significantly linked to well-being, suggesting that access alone does not guarantee engagement with nature.

Key Strategies

Recognizing the importance of equitable access to green spaces for resident engagement with nature, Singapore has integrated park provision into its urban strategy using two key planning parameters. The first is the park provision ratio—the amount of park area per 1,000 residents. This standard increased from 0.13 hectares per 1,000 residents in 1971 to 0.36 hectares in 1977, and has been maintained at 0.8 hectares since 1989, guiding the planning and expansion of parks across the city-state (Tan et al. 2013). The second parameter ensures that every household is within a 10-minute walk of a park, supporting both accessibility and inclusivity in urban green space planning (Er 2021).

Central to these efforts is a network of more than 300 parks, tiered by size to meet both neighborhood and city-wide needs. This is complemented by an extensive system of over 390 km of park connectors, providing direct access from residential areas and significantly enhancing green space connectivity across the island (Er and Leong 2023).

- **Facilitating Nature Interaction through Landscape Design and Horticultural Programs**

  While all green spaces benefit human well-being, certain features within natural settings—such as varied landforms, rich biodiversity, and tranquility—are especially effective at enhancing mental health (Olszewska et al. 2016). The Contemplative Landscape Model (CLM) offers an evidence-based framework for designing outdoor environments that promote reflection and relaxation. In Singapore, applying CLM principles has proven effective; parks with increased biodiversity, layered vegetation, and a tranquil atmosphere scored highly for contemplative quality and supported psychological restoration (Olszewska-Guizzo et al. 2022).

  Another area of focus in Singapore has been the application of therapeutic horticulture, which is defined as a facilitated nature-based programme where participants engage in plant-related activities, such as gardening, nature-inspired art, or sensory experiences with plants-to enhance their well-being. Studies have shown that participation in therapeutic horticulture leads to significant improvements in affective well-being, including reduced anxiety, healthier sleep patterns, improved cognitive functioning, and increased happiness after each session (Sia et al. 2020).

  In high-density cities, where most residents lack access to private gardens, providing communal gardening spaces is essential for realizing the mental health benefits associated with gardening. Research in Singapore showed that gardening significantly improved mental resilience and well-being, reducing anxiety and enhancing mood (Sia et al. 2022).

In this context, community-driven urban greenery is an emerging and promising dimension of tropical urban ecology. Initiatives such as ground-level community gardens and informal plantings along residential corridors are often stewarded by local residents, with government facilitated efforts (Sia et al. 2023a, b). These forms of greening confer several ecological, social, and economic advantages. They reduce governance costs by decentralizing design and maintenance, allowing residents to select plants for local ecosystem services like food production. Beyond individual benefits, community gardens offer broader ecological and social value: they support greater native biodiversity when compared to efforts by commercial landscaping (Oh et al. 2018). They foster social cohesion by encouraging cooperative activities like planting and harvesting, which build trust and reciprocity among participants (Oh et al. 2022). These micro-interactions help counter urban isolation, enhancing both individual well-being and collective resilience (Soga et al. 2017). Thus, providing access to community gardens in urban environments not only addresses the lack of private green space but also promotes biodiversity and strengthens social bonds.

Key Strategies

The evidence-based design principles have been applied on both existing and new parks. Notable examples include the Learning Forest at the Singapore Botanic Gardens (Fig. 4), where landscape scenes were curated to elicit the characteristics of the CLM:

Fig. 4 Landscape scene with high CLM score at the Learning Forest, Singapore Botanic Gardens

- Layers of the Landscape—The vista exhibits a strong sense of depth of view with a distinct fore-, middle- and back-ground.
- Landform—Undulating lines, asymmetry and sky views dominate the scene.
- Biodiversity—The Learning Forest features a lush lowland rainforest and freshwater wetland ecosystem containing a rich biodiversity.
- Color & Light—The scene exhibits a play of light and shade making the landscape appear more animated and alive.
- Compatibility—The natural and man-made elements seamlessly integrate forming a balanced and harmonious composition.
- Archetypal Elements—The symbolic elements of reflective water, mature trees and forest backdrop are the dominating presence within the landscape.
- Character of Peace & Silence—The scene evokes a sense of solitude and relaxation away from the intensity of urban life.

The findings on effects of therapeutic horticulture support the development of therapeutic gardens—such as the example shown in Figure 5—which exemplify a thoughtful approach to landscape design that prioritizes user comfort and accessibility. Key features include dense shade from mature trees and natural ventilation, which create a cooler and more comfortable microclimate. Clear, well-connected layouts with wide, looped pathways ensure easy navigation and accessibility for all visitors, including those using wheelchairs. Curving paths and strategically placed vegetation add a sense of intrigue and encourage exploration. The gardens also provide rich sensory experiences through diverse plantings—fragrant, edible and colorful species that attract birds and butterflies. Ergonomically designed raised planters further enhance accessibility, making these gardens inclusive and engaging for everyone (National Parks Board 2017).

Fig. 5 Therapeutic Garden at HortPark

4.2 Comparative Perspectives: Singapore and Other Global Cities

To situate Singapore's standing in an international context, data was analyzed using two established measures: the Inclusion of Nature in Self (INS) scale for nature connectedness (Schultz 2001) and the WHO-5 Well-Being Index for mental well-being (World Health Organization 1998). The comparison of Singapore's scores with data from 18 international countries collected by White et al. (2021)—most of which have significantly lower population densities—found that Singapore exhibits comparatively high levels of both nature connectedness (4.03) and mental well-being (81.35) (Fig. 6).

This observation challenges the common assumption that highly urbanized environments inevitably reduce individuals' sense of connection to nature. Notably, both INS and WHO-5 scores for Singapore were considerably higher than those reported for Hong Kong (INS: 2.96; WHO-5: 53.02), despite both cities sharing similarly high population densities. These results corroborate earlier research which found no evidence of an "extinction of experience" among Singapore residents (Oh et al. 2020). Over a 20-year period, there was no significant decline in residents' interactions with nature or in their sense of connection to it. These findings suggest that intentional and comprehensive urban greening strategies, alongside targeted nature engagement programmes may buffer against the physical and emotional disconnection from nature often associated with urban living.

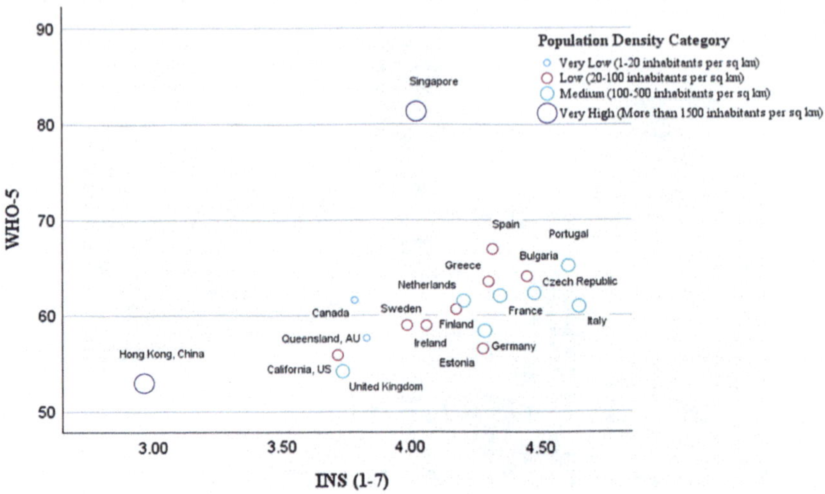

Fig. 6 Relationships between positive wellbeing (WHO-5) and nature connectedness (INS). Population density categories: very low (1–20), low (20–100), medium (100–500), high (500–1,500) and very high (>1,500) people per km^2. (Adapted from White et al. 2021)

5 Integrating Nature-Based Solutions in Urban Landscapes

This chapter underscores critical imperatives to integrate nature-based solutions into tropical urban landscapes. To cultivate nature connectedness and thereby promote well-being, spatial strategies must prioritize proximity, diversity (a connected network of managed parks, nature parks, and nature reserves), and biophilic design, including the creation of therapeutic landscapes with restorative potential. These principles enhance the quality, frequency and duration of human-nature interactions. They support biodiversity, while improving human health through regular, sustained, and meaningful engagement with nature.

Singapore provides a compelling case study in balancing urban density with meaningful nature interaction. It has achieved a model of ecological integration through ecologically rich, multifunctional urban green spaces for both accessibility and quality of experience. The city's integration of interconnected green networks and restoration of natural habitats exemplifies how tropical cities can create the physical conditions that both foster nature connectedness and conserve native biodiversity.

For other tropical cities, Singapore's experiences indicate that adopting multi-level, culturally relevant nature-based solutions—rather than isolated green spaces—could help advance social inclusivity, and overall well-being. This integrated approach may offer synergistic benefits, as improving everyday access to nature not only supports public health but also contributes to climate adaptation and biodiversity goals (McDonald et al. 2018). Reimagining urban nature as an interconnected system might be especially valuable for tropical cities facing the dual challenges of rapid urbanization and environmental change.

References

Bashan D, Colléony A, Shwartz A (2021) Urban versus rural? The effects of residential status on species identification skills and connection to nature. People Nat 3(2):347–358. https://doi.org/10.1002/pan3.10176

Beatley T (2011) Biophilic cities: integrating nature into urban design and planning. Island Press.

Belcher RN, Chisholm RA (2018) Tropical vegetation and residential property value: a hedonic pricing analysis in Singapore. Ecol Econ 149:149–159. https://doi.org/10.1016/j.ecolecon.2018.03.012

Berdejo-Espinola V, Suárez-Castro AF, Amano T, Fielding KS, Oh RR, Fuller RA (2021) Urban Green space use during a time of stress: a case study during the Covid-19 pandemic in Brisbane, Australia. People Nat 3(3):597–609. https://doi.org/10.1002/pan3.10218

Bonebrake TC, Tsang TP, Yu N, Wang Y, Ledger MJ, Tilley HB, Yau EY, Andersson AA, Boyle MJ, Lee KW, Li Q, Ling YF, Dongmo MA, Güçlü C, Dingle C, Ashton LA (2024) Tropical cities as windows into the ecosystems of our present and future. Biotropica 57(1). https://doi.org/10.1111/btp.13369

Cardinali M, Beenackers MA, Fleury-Bahi G, Bodénan P, Petrova MT, van Timmeren A, Pottgiesser U (2024) Examining green space characteristics for social cohesion and mental health outcomes: a sensitivity analysis in four European cities. Urban Forest Urban Greening 93:128230. https://doi.org/10.1016/j.ufug.2024.128230

Centre For Liveable Cities, Singapore, National Parks Board (Singapore, & Cengage Learning (2015) Biodiversity: nature conservation in the greening of Singapore. Cengage Learning. National Parks Board. (2017). Design guidelines for therapeutic gardens in Singapore. https://www.nparks.gov.sg/~/media/nparks-real-content/gardens-parks-and-nature/therapeutic-gardens/designguidelines_for_therapeuticgardens_in_sg.pdf

Chang C, Oh RR, Nghiem TP, Zhang Y, Tan CLY, Lin BB, Gaston KJ, Fuller RA, Carrasco LR (2020) Life satisfaction linked to the diversity of nature experiences and nature views from the window. Landsc Urban Plan 202:103874. https://doi.org/10.1016/j.landurbplan.2020.103874

Chisholm RA, Kristensen NP, Rheindt FE, Chong KY, Ascher JS, Lim KK, Ng PK, Yeo DC, Meier R, Tan HH, Giam X, Yeoh YS, Seah WW, Berman LM, Tan HZ, Sadanandan KR, Theng M, Jusoh WF, Jain A, Sin YC et al (2023) Two centuries of biodiversity discovery and loss in Singapore. Proc Natl Acad Sci 120(51). https://doi.org/10.1073/pnas.2309034120

Deng Z, Villarini G, Wang Z (2025) Climate change dominates over urbanization in tropical cyclone rainfall patterns. Commun Earth Environ 6(1). https://doi.org/10.1038/s43247-025-02048-z

Department of Statistics Singapore. (2019). Singapore in figures 2019

Engemann K, Pedersen CB, Arge L, Tsirogiannis C, Mortensen PB, Svenning J-C (2019) Residential green space in childhood is associated with lower risk of psychiatric disorders from adolescence into adulthood. Proc Natl Acad Sci 116(11):5188–5193. https://doi.org/10.1073/pnas.1807504116

Er K (2021) Transforming Singapore into a City in Nature. https://www.clc.gov.sg/docs/default-source/urban-solutions/urbsol19pdf/09_essay_transforming-singapore-into-a-city-in-nature.pdf

Er K, Leong CC (2023) Greening Singapore: from garden city to city in nature. In: Peace with Nature. Essay, World Scientific, pp 79–86

Fuller RA, Irvine KN, Devine-Wright P, Warren PH, Gaston KJ (2007) Psychological benefits of greenspace increase with biodiversity. Biol Let 3(4):390–394. https://doi.org/10.1098/rsbl.2007.0149

Louv R (2005) Last child in the woods: saving our children from nature deficit disorder. Algonquin Books

Marvier M, Kareiva P, Felix D, Ferrante BJ, Billington MB (2023) The benefits of nature exposure: the need for research that better informs implementation. Proc Natl Acad Sci 120(44). https://doi.org/10.1073/pnas.2304126120

Mayer FS, Frantz CM (2004) The connectedness to nature scale: a measure of individuals' feeling in community with nature. J Environ Psychol 24(4):503–515. https://doi.org/10.1016/j.jenvp.2004.10.001

McDonald RI, Colbert M, Hamann M, Simkin R, Walsh B (Eds) (2018) Nature in the urban century: a global assessment of where and how to conserve nature for biodiversity and human well-being. The Nature Conservancy, Arlington, VA. www.nature.org/content/dam/tnc/nature/en/documents/TNC_NatureintheUrbanCentury_FullReport.pdf

National Parks Board (2017) Design guidelines for therapeutic gardens in Singapore. https://www.nparks.gov.sg/~/media/nparks-real-content/gardens-parks-and-nature/therapeutic-gardens/designguidelines_for_therapeuticgardens_in_sg.pdf

Nghiem LTP, Zhang Y, Oh RR, Chang C, Tan CLY, Shannahan DF, Lin BB, Gaston KJ, Fuller RA, Carrasco LR (2021a) Equity in green and blue spaces availability in Singapore. Landsc Urban Plan 210:104083. https://doi.org/10.1016/j.landurbplan.2021.104083

Nghiem TPL, Wong KL, Jeevanandam L, Chang CC, Tan LYC, Goh Y, Carrasco LR (2021) Biodiverse urban forests, happy people: experimental evidence linking perceived biodiversity, restoration, and emotional wellbeing. Urban Forest Urban Greening 59:127030. https://doi.org/10.1016/j.ufug.2021.127030

Nisbet EK, Zelenski JM, Murphy SA (2008) The nature relatedness scale. Environ Behav 41(5):715–740. https://doi.org/10.1177/0013916508318748

Oh RRY, Fielding KS, Nghiem TP, Chang C, Shanahan DF, Gaston KJ, Carrasco RL, Fuller RA (2021a) Factors influencing nature interactions vary between cities and types of nature interactions. People Nat 3(2):405–417. https://doi.org/10.1002/pan3.10181

Oh RRY, Fielding KS, Chang C-C, Nghiem LT, Tan CL, Quazi SA, Shanahan DF, Gaston KJ, Carrasco RL, Fuller RA (2021b) Health and wellbeing benefits from nature experiences in tropical settings depend on strength of connection to nature. Int J Environ Res Public Health 18(19):10149. https://doi.org/10.3390/ijerph181910149

Oh RRY, Fielding KS, Carrasco RL, Fuller RA (2020) No evidence of an extinction of experience or emotional disconnect from nature in Urban Singapore. People Nat 2(4):1196–1209. https://doi.org/10.1002/pan3.10148

Oh RRY, Richards DR, Yee ATK (2018) Community-driven skyrise greenery in a dense tropical city provides biodiversity and ecosystem service benefits. Landsc Urban Plan 169:115–123. https://doi.org/10.1016/j.landurbplan.2017.08.014

Oh RRY, Zhang Y, Nghiem LTP, Chang C, Tan CLY, Quazi SA, Shanahan DF, Lin BB, Gaston KJ, Fuller RA, Carrasco RL (2022) Connection to nature and time spent in gardens predicts social cohesion. Urban Forest Urban Greening 74:127655. https://doi.org/10.1016/j.ufug.2022.127655

Olszewska AA, Marques PF, Ryan RL, Barbosa F (2016) What makes a landscape contemplative? Environ Planning B: Urban Anal City Sci 45(1):7–25. https://doi.org/10.1177/0265813516660716

Olszewska-Guizzo A, Sia A, Fogel A, Ho R (2022) Features of urban green spaces associated with positive emotions, mindfulness and relaxation. Sci Rep 12(1). https://doi.org/10.1038/s41598-022-24637-0

Petrunoff NA, Yi NX, Dickens B, Sia A, Koo J, Cook AR, Lin WH, Ying L, Hsing AW, van Dam RM, Müller-Riemenschneider F (2021) Associations of park access, park use and physical activity in parks with wellbeing in an Asian urban environment: a cross-sectional study. *International* J Behav Nutr Phys Act 18(1). https://doi.org/10.1186/s12966-021-01147-2

Pyle RM (1978) The extinction of experience. Horticulture 56:64–67

Roth M (2007) Review of urban climate research in (sub)tropical regions. Int J Climatol 27(14):1859–1873. https://doi.org/10.1002/joc.1591

Sarkar C, Webster C, Gallacher J (2018) Residential greenness and prevalence of major depressive disorders: a cross-sectional, observational, associational study of 94 879 adult UK Biobank participants. The Lancet Planetary Health 2(4). https://doi.org/10.1016/s2542-5196(18)30051-2

Schultz PW (2001) The structure of environmental concern: concern for self, other people, and the biosphere. J Environ Psychol 21(4):327–339. https://doi.org/10.1006/jevp.2001.0227

Shanahan DF, Fuller RA, Bush R, Lin BB, Gaston KJ (2015) The health benefits of urban nature: how much do we need? Bioscience 65(5):476–485. https://doi.org/10.1093/biosci/biv032

Sia A, Soh MCK, Lee WM, Poon HJ, Er KBH (2025) Well-being of park users in a tropical city may be more influenced by their nature connectedness and exposure than bird diversity. Urban Forest Urban Greening 107:128806. https://doi.org/10.1016/j.ufug.2025.128806

Sia A, Tam WW, Fogel A, Kua EH, Khoo K, Ho RC (2020) Nature-based activities improve the well-being of older adults. Sci Rep 10(1). https://doi.org/10.1038/s41598-020-74828-w

Sia A, Tan PY, Er KB (2023a) The contributions of urban horticulture to cities' liveability and resilience: insights from Singapore. Plants, People, Planet 5(6):828–841. https://doi.org/10.1002/ppp3.10377

Sia A, Tan PY, Kim YJ, Er KB (2023b) Use and non-use of parks are dictated by nature orientation, perceived accessibility and social norm which manifest in a continuum. Landsc Urban Plan 235:104758. https://doi.org/10.1016/j.landurbplan.2023.104758

Sia A, Tan PY, Wong JC, Araib S, Ang WF, Er KB (2022) The impact of gardening on mental resilience in times of stress: a case study during the COVID-19 pandemic in Singapore. Urban Forest Urban Greening 68:127448. https://doi.org/10.1016/j.ufug.2021.127448

Simons G, Baldwin DS (2021) A critical review of the definition of 'wellbeing' for doctors and their patients in a post covid-19 era. Int J Soc Psychiatry 67(8):984–991. https://doi.org/10.1177/00207640211032259

Soga M, Gaston KJ (2016) Extinction of experience: the loss of human–nature interactions. Front Ecol Environ 14(2):94–101. https://doi.org/10.1002/fee.1225

Soga M, Gaston KJ (2023) Global synthesis reveals heterogeneous changes in connection of humans to nature. One Earth 6(2):131–138. https://doi.org/10.1016/j.oneear.2023.01.007

Soga M, Gaston KJ, Yamaura Y (2017) Gardening is beneficial for health: a meta-analysis. Preventive Medicine Reports 5:92–99. https://doi.org/10.1016/j.pmedr.2016.11.007

State of the Tropics (2020) State of the tropics 2020 report. James Cook University

Sustainable Singapore Blueprint (2015) Sustainable Singapore blueprint 2015. Ministry of the Environment and Water Resources and Ministry of National Development, Singapore

Tan PY, Wang J, Sia A (2013) Perspectives on five decades of the urban greening of Singapore. Cities 32:24–32. https://doi.org/10.1016/j.cities.2013.02.001

White MP, Elliott LR, Grellier J, Economou T, Bell S, Bratman GN, Cirach M, Gascon M, Lima ML, Lõhmus M, Nieuwenhuijsen M, Ojala A, Roiko A, Schultz PW, van den Bosch M, Fleming LE (2021) Associations between green/blue spaces and mental health across 18 countries. Sci Rep 11(1). https://doi.org/10.1038/s41598-021-87675-0

White MP, Hartig T, Martin L, Pahl S, van den Berg AE, Wells NM, Costongs C, Dzhambov AM, Elliott LR, Godfrey A, Hartl A, Konijnendijk C, Litt JS, Lovell R, Lymeus F, O'Driscoll C, Pichler C, Pouso S, Razani N, van den Bosch M et al (2023) Nature-based biopsychosocial resilience: an integrative theoretical framework for research on nature and health. Environ Int 181:108234. https://doi.org/10.1016/j.envint.2023.108234

World Health Organization. Regional Office for Europe (1998) Wellbeing measures in primary health care/the DepCare Project: report on a WHO meeting: Stockholm, Sweden, 12–13 February 1998

Yap KK, Soh MC, Sia A, Chin WJ, Araib S, Ang WP, Tan PY, Er KB (2022) The influence of the covid-19 pandemic on the demand for different shades of green. People Nat 4(2):505–518. https://doi.org/10.1002/pan3.10304

Climate Resilience and Urban Cooling

Explore the Application of Green Infrastructure in Enhancing Climate Resilience in China: A Comparative Analysis in Suzhou Industrial Park (SIP) and Gusu District, Suzhou

Yanhui Lei, Ziyu Fan, and Sifan Yu

Abstract Traditional urban construction often prioritizes grey infrastructure while neglecting the natural functions of ecosystems, exacerbating climate change issues. In response, green infrastructure (GI) has been developed and adapted as a nature-based solution (NBS) and has been proven to provide various ecosystem services including reducing flood risk and mitigating the climate change, which contributes to urban resilience. Currently, most Chinese cities still rely on grey infrastructure, and relevant studies remain limited and usually substitute the term 'sponge city', which can lead to a conceptual deviation from the concept of GI. Additionally, there is a lack of comparative research on the application of GI in areas with different development conditions within cities, such as newly developed districts versus historic urban areas. To fill in this gap, this study employs a combined qualitative and quantitative method that covers literature review, ArcGIS analysis, and survey questionnaires to examine the adaptability and effectiveness of GI in terms of stormwater management in newly developed district (Suzhou Industrial Park) and old district (Gusu District) in Suzhou. The expected outcomes could be the identification of GI distribution, categories, and effects in terms of stormwater management in the two districts. Finally, this paper proposes potential optimization strategies based on Suzhou's experience to other cities within the Yangtze River Delta.

Keywords Nature-based solutions · Adaptability of green infrastructure · Climate resilience · Suzhou Industrial Park · Gusu district

Y. Lei (✉) · Z. Fan · S. Yu
Department of Urban Planning and Design, Xi'an Jiaotong-Liverpool University, Suzhou, China
e-mail: Yanhui.Lei@xjtlu.edu.cn

A. Cheshmehzangi et al. (eds.), *Nature in Cities, Nurturing Cities*, Urban Sustainability,
https://doi.org/10.1007/978-981-95-5487-4_5

1 Introduction

1.1 Background

Under the global context of climate change, extreme rainfall events are occurring more frequently. Increasing area of impervious surface including pavements and rooftops is replacing vegetation and soil in cities (Vijayaraghavan et al. 2021). Coupled with factors such as the continuously increasing population and unplanned development in urban areas, the urban flooding management is receiving increasing attention and becoming a priority (Chen et al. 2021). In addition, traditional urban construction often prioritizes grey infrastructure while neglecting the natural functions of ecosystems. With the rapid urbanization and more frequent extreme rainfall events, traditional grey infrastructure has become not so effective and efficient any more (Xu et al. 2019). In the last few decades, most Chinese cities used hard-engineering control measures including dams and impoundments in response to urban flooding issues, and currently still mainly rely on grey infrastructure for drainage system. Such approach of mitigating flood intensity through enhancing flood division could be effective for upstream catchment areas, while cannot guarantee the protection of downstream areas (Chan et al. 2018). Moreover, such grey infrastructure is mainly composed of concrete and steel, whose production can be a burden on the environment (Vineyard et al. 2015).

In response, green infrastructure (GI), a nature-based solution that was initially proposed in America in the 1990s (Fletcher et al. 2015), has since evolved and adaptively applied as a complement or even an alternative to traditional grey drainage system (Chen et al. 2021; Vineyard et al. 2015). In current studies, GI is mainly defined as a term for urban stormwater management (Berland et al. 2017; Chini et al. 2017; Fletcher et al. 2015; Nguyen et al. 2019), as well as a network of green spaces composed of natural, semi-natural, and artificial ecosystems (Matsler et al. 2021; Nieuwenhuijsen 2025; Tzoulas et al. 2007; Ying et al. 2022). Moreover, the application of GI is considered as an effective and flexible strategy for dealing with issues raised by climate change and urbanization (Dong et al. 2017), and has been proven to provide various ecosystem services, including reducing flood risks (Berland et al. 2017; Chini et al. 2017; Fletcher et al. 2015; Nguyen et al. 2019; Semeraro et al. 2021), and mitigating urban heat island effects (Semeraro et al. 2021; Tzoulas et al. 2007; Ying et al. 2022).

Despite the studies on GI increasing, there are few focused on China (Ying et al. 2022). Furthermore, relevant research in China often uses 'sponge city' as an alternative term (Lei 2024) which raises a deviation from GI. Additionally, existing cases mostly focus on the macro scale and lack comparative analyses of districts with different development situations within a city. Newly developed districts can adopt rational masterplan in advance to support the implementation of GI, while it is often constrained by limited land resources and historical preservation requirements in old districts. Therefore, the adaptive application of GI in regions with different development conditions should be further studied to promote sustainable development.

1.2 Research Aim and Research Question

Under the context, this research aims to systematically analyze the differences in the adaptability and effectiveness of GI in the newly developed district, Suzhou Industrial Park (SIP), and old district, Gusu District (GD) of Suzhou, which is a city in the Yangtze River Delta that owns abundant water resources while also faces the practical challenges of climate adaptation and rainwater environment management as well. Additionally, its spatial structure is featured by coexistence of newly developed district and old district, providing proper comparative cases. In 2016, Suzhou became a pilot city of 'sponge city', having accumulated initial experience. Therefore, GI has great potential and necessity for implementation in Suzhou, which deserves further research. In this study, following research questions are mainly focused:

1. What are the differences of the current state of GI implementation in SIP and GD?
2. What are the differences in the effectiveness of GI implementation in SIP and GD?
3. How can the experience be applied to other Yangtze River Delta cities for the construction of GI?

2 Literature Review

2.1 Concepts of GI

In different studies, GI is usually conceptualized in various ways, with differing emphases depending on disciplinary perspectives. Broadly, the concepts of GI can be mainly classified into three categories, which are greenspace planning, urban ecology, and stormwater management concept (Matsler et al. 2021). In terms of greenspace planning, GI is recognized as a spatial strategy for organizing the urban environment that emphasizes the spatial layout of green spaces within and around urban areas, which can support a range of ecological and cultural functional concerns, such as providing social and environmental benefits through green corridors or green belts (Ahern 2007; Matsler et al. 2021; Mell 2008). In addition, it also focuses on the interactions between different types of urban ecosystems and built infrastructures for socially negotiated goals (Grabowski et al. 2022).

For urban ecology concept, GI is usually described and analyzed as a multifunctional network of vegetation in urban area (Matsler et al. 2021). In 2013, the European Commission defined GI as a network of natural and semi-natural areas which aims to provide a broad range of ecosystem services (Addo-Bankas et al. 2024; Hansen and Pauleit 2014; Monteiro et al. 2020). Under this concept, studies tend to be concerned with the ecological functions and ecosystem services provided by different natural systems in the urban environment.

Regarding the concept of stormwater management, although it is considered somewhat narrowly defined as it usually focuses on designs that improve hydrological function and may not even include vegetation (Matsler et al. 2021), it remains a widely adopted and actively studied dimension of GI research. Ying et al. (2022) used CiteSpace and VOSviewer to review and visualize the core collection of the Web of Science database with the key word 'green infrastructure', finding that the stormwater management has the highest frequency and represented one of the core research directions related to GI. Moreover, GI under this concept has been evolved into locally specific concepts or strategies in different countries, such as the low impact development (LID), low impact urban design and development (LIUDD) and the best management practices (BMPs) in North America (Ayoubi Ayoublu et al. 2024; Fletcher et al. 2015; Li et al. 2016; Liu et al. 2017, 2021; Xia et al. 2017; Zhang et al. 2019), the water sensitive urban design (WSUD) in Australia (Ferrans et al. 2022; Fletcher et al. 2015; Lashford et al. 2019; Seyedashraf et al. 2022; Xia et al. 2017), the sustainable urban drainage systems (SUDS) in the UK (Beecham et al. 2018; Li et al. 2016; Xia et al. 2017), and the sponge city (SC) in China (Chan et al. 2018; Jiang et al. 2018; Liu et al. 2017; Wang et al. 2018).

Generally, GI represents an integrated approach to environmental, social, and economic challenges as an essential framework in contemporary urban and landscape planning. Given the multidisciplinary nature of GI and its varying emphases across different contexts, no single definition can fully capture its complexity (Addo-Bankas et al. 2024; Matsler et al. 2021; Seiwert and Rößler 2020). Furthermore, since each conceptualization of GI has different scales, functions, disciplinary orientations, it can lead to great difference in terms of implementations and the delivery of benefits (Matsler et al. 2021). Wang and Banzhaf (2018) also pointed out that it is possible to make GI vague and ambiguous when using certain terms and definitions in place of GI without conceptual assumptions or specific distinctions, which would undermine its effectiveness, multifunctionality, and potential for innovation. Consequently, it is essential to clearly define the objects and scales of GI so that can ensure the accuracy, effectiveness, and applicability of research results and practices. In this study, GI mainly refers to a kind of stormwater management measure in urban areas that are centered by green elements including vegetations, aiming to regulate, purify, and infiltrate stormwater through natural processes, thereby enhancing the resilience to climatic and hydrological changes.

2.2 *Classification and Element Identification of GI*

Given the diversity of disciplines, contexts, terminologies, aims, and assessment criteria, there is no consensus on how GI should be classified. Moreover, it is unlikely that a universal classification approach applicable to all contexts can be established (Bartesaghi Koc et al. 2017). In response, researchers try to develop and use typologies that can transform complex and diverse green space data information into clear GI categories. In addition, Ignatieva and Mofrad (2023) suggested that GI should be categorized based on the local context of the study areas, which is followed by many studies in different fields. In the study on the relationship between GI and human health, Suppakittpaisarn et al. (2017) simply divided GI into trees, green spaces, and other types of GI including green roofs, rain gardens, and bioswales, due to the limited number of relevant studies available. Such broad categories can be sufficient in the field, while it lacks the granularity required for practical applications in urban planning.

Building on this issue of classification precision, Ignatieva and Mofrad (2023) pointed out that one of the most frequently applied green space typologies is based on the extent of natural characteristics. In their research on the ecological contributions of green spaces to GI in Canberra, they followed such approach to divide green spaces into natural, semi-natural, and human-made landscapes. Furthermore, specific green space types were subdivided based on the site conditions and vegetation types of the study area, which cover woodlands, parks, and residential green spaces. Similarly, Hanna et al. (2024) also applied the classification approach when examining the ecosystem services that GI can provide in Zaragoza, and additionally included rivers and forests in the peri-urban areas as GI elements due to the differences in study sites.

From a climatological perspective, Koc et al. (2016) proposed a more detailed and standardized classification scheme for identifying and characterizing GI assets. It classified GI features into three main categories according to the spatial structure: vegetation layers, ground surfaces, and building structures. These categories are further subdivided based on factors such as vegetation height and permeability, and are combined through a dual-input matrix, which is then divided into four main categories: tree canopy, green open spaces, green roofs, and vertical greenery systems. This classification approach offers higher spatial identification accuracy and describes the detailed features of various GI assets. However, its complexity and reliance on the accuracy of remote sensing data may make it difficult to implement in urban planning practice. On the scheme basis, Koc et al. (2017) identified and reviewed literature on GI assets and components from 15 countries and regions, including the UK, Germany, and Spain. They pointed out that existing studies often classify green spaces based on the functionality, structural morphology, and configuration of GI. From a structural-configurational perspective, the studies that employ a combination methodology of literature review, case studies, and GIS analysis typically use land-use land-cover (LULC) to segment GI into more detailed categories.

Based on the above synthesis, for the purpose of subsequent analyses on the distribution of GI in urban areas and its effectiveness in stormwater management, this study also classifies GI into natural, semi-natural, and human-made categories according to the degree of naturalness. In addition, within the context of the GI concept and study area in this study, natural wetlands, natural woodlands, artificial wetlands, urban street trees, shrub planting belts, grassland, agricultural croplands, rain gardens, green roofs, bioretention, and vegetated permeable pavements will be explored as the main GI elements in the following literature review section.

2.3 Stormwater Management Effectiveness Framework Based on GI

When examining the functions and effectiveness of GI in terms of stormwater management, researchers commonly assess its performance across multiple dimensions, including infiltration, retention, storage, purification, utilization, and discharge. For example, Wang et al. (2018) reviewed and synthesized key GI facilities, including permeable pavements, green roofs, rain gardens, and bioretention systems, focusing on their roles in stormwater utilization, groundwater recharge, peak flow reduction, purification, and conveyance of runoff. The study found that rain gardens consistently delivered above-average performance in reducing peak flows and improving water quality. Permeable pavements and bioretention systems were particularly effective in supporting groundwater recharge, highlighting their value in stormwater retention. In comparison, green roofs showed relatively modest benefits across these functions. Similarly, Jones et al. (2022) also evaluated various GI types in relation to the ecosystem services they provide. In the context of stormwater management, the study focused on their effectiveness in enhancing water quality and regulating hydrological flows. The findings suggest that wetlands, riparian woodlands, permeable surfaces, and grasslands are particularly beneficial for improving water quality. For water flow management, rain gardens, permeable pavements, green roofs, and areas with trees or shrubs performed well, while grasslands showed variable outcomes depending on soil compaction, with overall moderate-to-low range performance.

Woodlands, grasslands, shrub planting belts, and street trees, as main components of urban vegetation, show different advantages in stormwater infiltration, retention, storage, and purification. Grasslands are effective in reducing runoff, particularly by enhancing soil permeability. However, the hydrophobic soils can lead to rapid infiltration of water and pollutants, limiting their capacity for pollutant retention (Milazzo et al. 2023). In contrast, trees in both woodlands and streets can reduce runoff through multiple mechanisms such as stormwater interception, transpiration, and stemflow (Seitz and Escobedo 2008). In addition, trees help minimize inflows into stormwater infrastructure by capturing rainfall in the canopy, promoting infiltration through the root zone, and absorbing soil moisture, which collectively lowers treatment costs and pollutant transport risks (Dowtin et al. 2023). Tree roots can also enhance soil aggregation, support groundwater recharge, and help regulate soil moisture after storms through transpiration, thereby reducing runoff from subsequent rainfall events (Berland et al. 2017). Moreover, urban street trees tend to develop larger canopies due to more open space and reduced competition, which increases their interception capacity (Selbig et al. 2022). Shrub planting belts can slow surface flow by reducing raindrop impact and surface erosion as they have dense foliage, while their roots could lead to pollutant uptake and retention (Denman et al. 2011).

In addition to terrestrial vegetation, wetland ecosystem, both natural and artificial wetlands, play a vital role in mitigating floods and improving water quality. Most wetlands can effectively reduce or delay peak flows by storing excess water and gradually releasing it. This process helps to reduce downstream flood risks and supports groundwater recharge (Bullock and Acreman 2003). Furthermore, wetlands can also improve water quality by filtering suspended solids and removing, recycling, or immobilizing pollutants and nutrients. In addition, the vegetation diversity and expansive water surfaces help regulate flow, and plant root systems not only filter particulates but also contribute to nitrogen and phosphorus removal (Rooney et al. 2015). Small wetlands in upper catchments can delay flood peaks and reduce flow intensity through localized storage. Larger wetlands in lower catchments regulate water discharge and help lower flood levels during major storm events (Zedler 2003). However, there are also some limitations of natural wetlands in terms of purifying the stormwater. According to Zedler (2003), microbial processes can reduce nitrogen through transformation, but phosphorus and sediment retention remain less effective. In comparison, artificial wetlands have better performance in pollutant removal (Wong et al. 1999), especially in controlling nitrogen and phosphorus in stormwater (Vymazal 2007). Although artificial wetlands have higher pollutant removal efficiency, they tend to be inferior to natural wetlands in terms of sustainable nutrient control and water quality stability (Rooney et al. 2015).

Agricultural cropland plays a significant role in combating climate change, mitigating floods, and protecting water quality, while it is less studied as a form of GI. As a naturally permeable surface, croplands can slow surface runoff and store stormwater, helping to reduce flood risk and support groundwater recharge (Lee and Liu 2023). Moreover, the use of rice paddy bunds can further enhance its stormwater retention capacity (Lee and Liu 2023). Due to the well-structured soils and continuous vegetation cover, agricultural croplands can not only facilitate stormwater interception, and infiltration, but also reduce the runoff and erosion (Power 2010).

Among designed GI facilities and systems, rain gardens have been extensively studied as they can effectively reduce the urban flooding risk and improve water quality by increasing stormwater infiltration capacity and slowing runoff peaks. The design of rain gardens typically includes shallow depressions and multiple layers of permeable media, which greatly increase the infiltration capacity of stormwater and can improve the quality of urban water environments by trapping pollutants in the initial rainfall, such as oils, heavy metals, and nutrients, and preventing them from entering the water body (Ishimatsu et al. 2017). Moreover, Tang et al. (2016) presented a four-year monitoring hydrological model and used it to study the stormwater retention performance of a rain garden in the semi-humid loess area of Xi'an. The results show that the rain garden has significant retention and reduction effects on stormwater, and has stable control benefits on longer time scales. However, they may become saturated and overflow during short periods of high intensity rainfall (Tang et al. 2016; Zhang et al. 2020). In addition, rain gardens are capable of temporarily storing stormwater and improving the quality of water flowing into neighboring water bodies through plant uptake and soil filtration (Malaviya et al. 2019; Sharma and Malaviya 2021). Although rain gardens can effectively remove nitrogen and sediment, they face challenges in controlling phosphorus pollution, mainly due to surface dry sediments and short hydraulic retention times (Xing et al. 2021).

Green roofs represent another widely adopted GI strategy with demonstrated benefits in stormwater regulation and pollutant reduction. By retaining water in soil and vegetation layers, green roofs delay runoff generation and reduce total runoff volume, easing pressure on urban drainage networks during storm events (Raimondi and Becciu 2021; Shafique et al. 2018; Stovin 2010; Vogel and Moore 2016). Moreover, many studies show that green roofs have a good purifying effect and can significantly reduce pollutant concentration especially when the system is stabilized (Razzaghmanesh et al. 2014). However, their effectiveness would decrease under sustained heavy rainfall or saturated conditions (Stovin 2010). In addition, compared to other GI systems, green roofs are generally less efficient in removing nitrogen and phosphorus, largely due to limited retention time (Xing et al. 2021).

Bioretention, as one of the most common GI interventions, significantly reduces runoff and enhances water quality. Bonciarelli et al. (2025) reviewed literature in the past decade, and concluded that bioretention systems can effectively slow peak flows and reduce runoff through infiltration, exfiltration, and evapotranspiration processes, particularly during small rainfall events. In addition, such systems can also remove suspended solids, heavy metals, and microplastics (Vogel and Moore 2016). The combined action of plant roots and soil media contributes to effective pollutant filtration (Nazarpour et al. 2023; Trowsdale and Simcock 2011). Nevertheless, during high-intensity rainfall, their peak flow reduction capacity may decline, and overflow risks increase under longer return-period storm events (Bonciarelli et al. 2025; Yang and Chui 2018). While BR systems perform well under moderate precipitation, their resilience during extreme weather events warrants further improvement.

Vegetated permeable pavements, integrating grassed swales with permeable surfaces, exhibit a more significant permeability than either grass-planted swales or permeable paving as a single system (Xie et al. 2017). Booth and Leavitt (1999) examined four permeable paving systems, which include one pavement with grass, and found that these permeable paving systems can almost eliminate surface runoff, significantly reduce peak stormwater flows, and delay stormwater catchment time. Moreover, studies have shown that permeable systems such as modular paving, grass grids, and gravel grids in the South Spanish region are able to reduce runoff and peak flows by more than 80% and extend runoff lag times by 70–80% (Rodak et al. 2019). Particularly, grass grid systems show better effectiveness in enhancing stormwater abatement through evapotranspiration by plants (Rodak et al. 2019). Generally, through the high infiltration capacity and the synergistic role of vegetation, vegetated permeable pavements can efficiently reduce surface runoff and enhance stormwater storage and purification, thereby improving water quality and mitigating flood risk (Vogel and Moore 2016).

Table 1 The functions and effectiveness of GI in study area

Main Classification	GI elements	Functions and effectiveness in stormwater management					
		Stormwater infiltration	Stormwater retention	Stormwater storage	Stormwater purification	Stormwater utilization	Stormwater discharge
Natural	Natural wetlands	Medium	High	High	Medium	Low	High
	Natural woodlands	High	Medium	Medium	Medium	Negligible	Medium
Semi-natural	Urban street trees	High	Medium	Low	Medium	Negligible	Medium
	Shrub planting belts	Medium	Low	Negligible	Medium	Negligible	Low
	Grassland	Medium	Low	Negligible	Low	Negligible	Low
	Agricultural croplands	High	Medium	Low	Low	Medium	Medium
Man-made	Artificial wetlands	Medium	High	High	High	Low	Medium
	Bioretention	High	Medium	Medium	High	Low	Medium
	Green roofs	Medium	Medium	Medium	Medium	Low	Low
	Rain gardens	High	Medium	Medium	Medium	Low	Medium
	Vegetated permeable pavements	High	Medium	Medium	Medium	Low	Medium

Source Made by authors

Overall, to synthesize the comparative findings and provide a concise overview of the stormwater management performance of various GI types, Table 1 summarizes their relative effectiveness across core functional criteria including infiltration, retention, storage, purification, utilization, and discharge.

3 Methodology

3.1 Research Design

This study employs the research methodology combined by qualitative and quantitative, which mainly includes literature review, data analyses based on GIS tools, and questionnaire surveys (Fig. 1). This case study compares the current spatial distribution and landscape patterns of SIP and GD GI based on GIS analysis of green cover and land cover data. In addition, the flood risk and GI mitigation effects of the two districts will be simulated and analyzed based on topography as well as precipitation data. Afterward, a questionnaire survey on the residents' evaluation of GI in SIP and

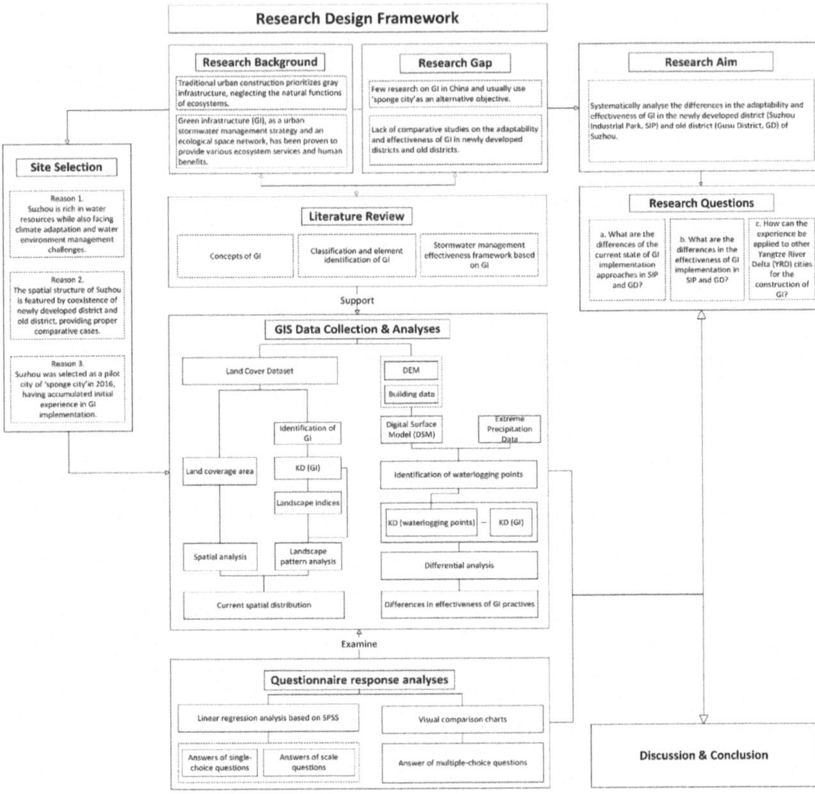

Fig. 1 Research design framework of the study (*Source* made by authors)

GD will be used to verify the GIS analyses. If needed, it can help to modify the final outcomes. Finally, it is expected to propose potentially applicable suggestions for other YRD cities according to the experience of Suzhou.

The Selection of the FRAGSTATS Indicators

To evaluate the spatial pattern differentiation of GI between GD and SIP, this study employed the SinoLC-1 1-m resolution land cover dataset and quantified landscape metrics using FRAGSTATS 4.3. Recognizing the limitations of single-parameter approaches in comprehensively interpreting landscape fragmentation processes (Davidson 1998), a multidimensional suite of indices was systematically selected: Patch Density, Mean Patch Shape Index, Mean Euclidean Nearest-Neighbor Distance, Contagion Index, Cohesion Index. The selection of these metrics aligns with established frameworks in landscape ecological analysis (Fan and Myint 2014; Nasehi and Namin 2020; Tian et al. 2011; Yang 2017), addressing both structural and functional dimensions of GI spatial organization.

3.2 Research Area

Suzhou (119°55′~121°2d0′ E, 30°47′~32°02′ N), a prefecture-level city in Jiangsu Province, is situated in the Yangtze River Delta region of eastern China (Fig. 2). Encompassing a total area of 8,657.32 km^2 with a permanent population of 12.958 million (Suzhou Municipal Bureau of Statistics 2024a, 2024b), it is classified as a megacity under China's State Council urban classification standards. The city's topography is characterized by flat terrain and an intricate network of waterways, emblematic of the Jiangnan water town archetype. Under the subtropical maritime monsoon climate regime, the region receives an average annual precipitation of 1,406.8 mm (Suzhou Municipal Bureau of Statistics 2024a), establishing favorable ecological foundations for green infrastructure development. This study specifically examines two administrative planning zones: Gusu District (GD) and Suzhou Industrial Park (SIP).

Gusu District (83.4 km^2), encompassing the UNESCO-protected Old Town, preserves the historic 'shuang qi pan' (double chessboard) urban configuration where waterways parallel street networks. Stringent height restrictions (≤24 m) and a preservation-centric spatial organization anchored by canal systems maintain the area's historical authenticity (Gusu District People's Government 2024). Constrained by limited land availability and heritage conservation policies, urban renewal in this district predominantly adopts a micro-intervention approach, with green spaces primarily concentrated in linear configurations along historic canals and alleyways.

Suzhou Industrial Park (278 km^2), established in 1994 adjacent to the eastern boundary of Gusu District, represents a strategic bilateral cooperation initiative

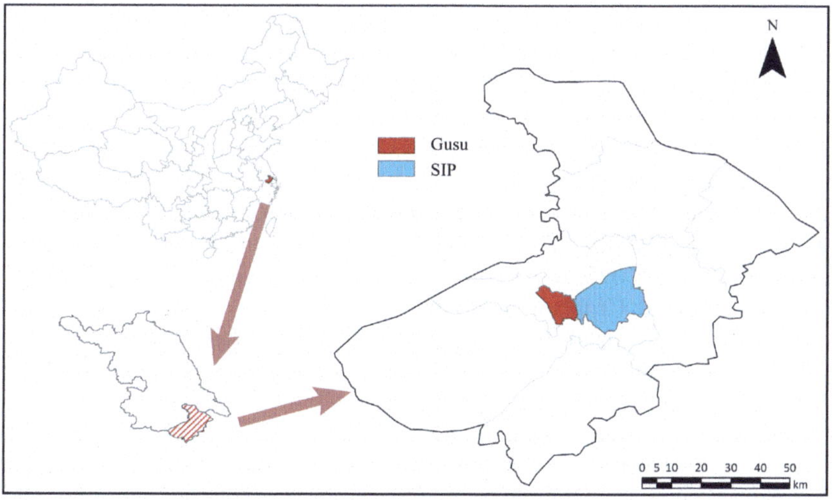

Fig. 2 Location of study area (*Source* Map made by the authors)

between the Chinese and Singaporean governments. Designed as a 'relatively self-contained industrial satellite town and garden city' under governmental planning directives, SIP contrasts markedly with Gusu District through systematic land consolidation and proactive planning reservations to implement urban green infrastructure, constructing a hierarchical green space system. Hydrological reconfiguration via lake reclamation transformed natural water bodies such as Jinji Lake and Dushu Lake into engineered ecological landscapes, which synergistically interact with modernist high-rise clusters to define the district's distinctive urban silhouette.

3.3 Data Collection

3.3.1 GIS Data

This study employs multi-source heterogeneous geospatial data, constructing an analytical database through systematic integration and preprocessing. Study area boundaries were acquired from authoritative national platforms and municipal government documentation: The administrative boundary of GD was derived from the administrative division vector dataset released by the National Geomatics Center of China on May 28, 2024. For SIP, due to dynamic administrative adjustments, its boundary was constructed through mosaicking of vector data published by the Suzhou Industrial Park Administrative Committee in 2024. Both datasets underwent coordinate system unification and topological correction to ensure spatial analytical accuracy.

Green Infrastructure (GI) spatial data were derived from the SinoLC-1 1-m resolution land cover dataset developed by the research team of Zhuohong Li and Wei He at Wuhan University (Li et al. 2023). As China's first nationwide 1-m resolution land cover product, this dataset leverages a deep learning framework (Low-to-High Framework, L2H-Frame) to harmonize multi-source open data, including global land cover products (GLC), OpenStreetMap (OSM), and Google Earth imagery (Li et al. 2023). It has been verified by the National Tibetan Plateau Data Center, with an overall classification accuracy of 73.61% and a Kappa coefficient of 0.6595 (Li et al. 2023). This study classified five land cover types from the SinoLC-1 1-m resolution dataset (Li et al. 2023) as GI components within the research area: Tree cover, Shrubland, Grassland, Cropland, and Wetland.

The terrain data utilized the ALOS PALSAR Global Digital Elevation Model (DEM, 12.5 m resolution 2011) provided by the Japan Aerospace Exploration Agency (JAXA). After acquiring the original data through NASA's EarthData platform, we constructed a high-precision 3D urban Digital Surface Model (DSM) by integrating building footprint vectors and height attributes from the AMAP Open Platform (2024). Specifically, an inverse distance weighting interpolation algorithm (IDW, power parameter $p = 2$) was applied to fuse ALOS DEM with 3D architectural data, generating a 1 m resolution DSM. This model accurately characterizes the combined effects of topographic relief and vertical building structures on surface hydrological processes.

The extreme precipitation data were obtained from the CHM_PRE China High-Resolution Precipitation Dataset (Version 2.1) developed by Hu and Miao (2023), hosted by the National Tibetan Plateau Data Center. Constructed using daily precipitation observations from 2,839 meteorological stations across China and adjacent regions spanning 1961 to present, this dataset employs an innovative 'precipitation background field + precipitation ratio field' methodology enhanced through monthly precipitation constraints and topographic feature corrections to optimize spatial interpolation accuracy (Hu and Miao 2023). The 0.1° resolution daily precipitation data were processed to extract decadal maximum daily precipitation records (2012–2022), from which extreme precipitation thresholds were calculated for GD and SIP. These thresholds were systematically integrated with the DSM to implement hydrological simulations for urban waterlogging hotspot identification.

3.3.2 Questionnaire

Based on a review of relevant literature and preliminary GIS analysis, this study employed a questionnaire to collect data on residents' everyday experiences and observation, which were subsequently used to verify and refine the GIS-based findings. The questionnaire was pilot-tested and revised prior to formal distribution. It comprised 17 questions addressing residents' perceptions of stormwater accumulation during rainy days in two selected areas, their understanding of green infrastructure (GI), and their routine observations of GI features. The design of the questionnaire was informed by literature and case study reviews, and it primarily consisted of single-choice, multiple-choice, and scale questions.

Firstly, it captured participants' demographic information, including age groups, place of residence, and time of residence. It then collected residents' feelings of rainfall and surface water accumulation and the extent to which these impacted their daily lives. In the section of understanding and feelings of GI, considering the variability in public awareness of GI, the questionnaire introduced eight common types of GI with brief descriptions and images for responders to better identify and make selections. The scale questions employed a 5-point Likert scale to assess perceived impacts of stormwater accumulation, perceived distribution of GI, and the perceived effectiveness of GI in mitigating daily stormwater ponding issue. A score of 1 represented 'no impact', 'very uneven', or 'not effective at all', and a score of 5 indicated 'very significant impact', 'very even', or 'very effective'. The study mainly used SPSS to firstly analyze the reliability of the Likert scale questions in the questionnaire, and then linear regression analyses were conducted on the responses to the single-choice and Likert scale questions to compare the differences between GD and SIP in terms of the perceived effects of stormwater ponding events and GI. The results of the multiple-choice questions would be analyzed in the discussion section and contribute to the recommendations for future improvements.

4 Results

4.1 *GIS Analyses*

4.1.1 Current Spatial Distribution

Spatial analysis of ArcGIS-processed SinoLC-1 land cover data (Table 2) reveals that SIP exhibits a marginally higher GI coverage ratio (20.30%) compared to GD (15.95%). Paradoxically, SIP demonstrates significantly greater impervious surface coverage, indicating heightened urbanization intensity and amplified surface runoff pressures. Conversely, GD's lower GI percentage coexists with proportionally higher waterbody coverage within its historic urban core. This configuration, integrating traditional canal systems with pocket green spaces, forms latent drainage networks that potentially enhance practical flood mitigation capacity through decentralized hydrological pathways.

The landscape pattern analysis reveals significant spatial distribution disparities in green infrastructure (GI) between GD and SIP (Table 3). GD exhibits higher values in patch density (PD) and mean Euclidean nearest-neighbor distance (ENN_MN) compared to SIP, indicating a highly fragmented GI configuration dominated by small, scattered patches. This spatial pattern aligns with GD's micro-intervention greening strategy, characterized by incremental upgrades of pocket green spaces within constrained urban fabrics. In contrast, SIP demonstrates markedly greater shape complexity (SHAPE_AM), reflecting intentionally designed ecological morphologies in its urban planning, such as expansive wetland parks and meandering green corridors.

Both districts show comparable landscape contagion (CONTAG), suggesting similar levels of spatial heterogeneity in GI distribution. However, the cohesion index (COHESION) reveals strong GI isolation in both areas, with spatial connectivity predominantly reliant on discrete, point-based green elements. Notably, despite SIP's advantages in morphological complexity and nominal connectivity metrics,

Table 2 Comparison of area by type in the two districts

AREA (sq m)	Green infrastructure	Impervious surface	Water body
GD	12,674,691.63	3,923,708.212	68,525,583.4
SIP	40,468,153.88	78,625,221.6	178,109,818.3

Table 3 Comparison of landscape indices between the two districts

Landscape indices	PD	SHAPE_AM	ENN_MN	CONTAG	COHESION
GD	723.4577	6.5814	5.3433	65.4997	99.9790
SIP	674.5937	9.0523	4.5842	65.3164	99.9801

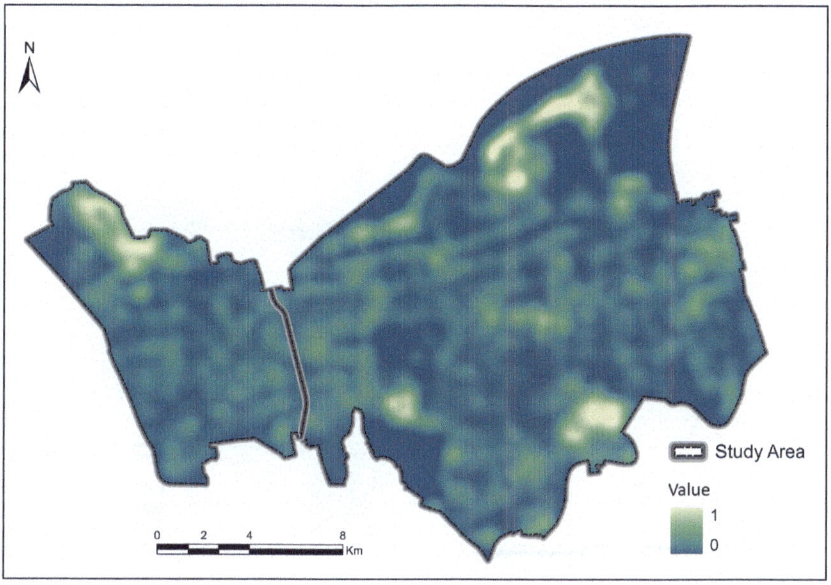

Fig. 3 Comparison of landscape indices between the two districts (*Source* Made by authors)

kernel density analysis (Fig. 3) identifies a core-periphery configuration—concentrated large green cores coexist with service-deficient gaps between industrial zones. This spatial mismatch results in localized deficiencies in runoff mitigation capacity.

4.1.2 Differences in Effectiveness of GI Practices

The study employed ArcGIS Pro to identify waterlogging hotspots under extreme precipitation scenarios through Digital Surface Model (DSM)-based hydrological modeling. Kernel density estimation (KDE) results for GD and SIP, normalized to a 0–1 scale, demonstrate significantly higher waterlogging risk in SIP compared to GD under identical topographic conditions (Fig. 4).

Figure 5 illustrates the differential analysis between waterlogging density and GI density, where negative values indicate effective GI-mediated risk mitigation, positive values denote risk amplification, and null values suggest either hydrological equilibrium or non-sensitive zones. GD exhibits predominantly negative values, whereas SIP shows mixed outcomes with persistent positive values across extensive areas, revealing spatial heterogeneity in GI functionality. The results may be due to the GI design capacity of SIP not accounting for extreme rainfall intensity under climate change, limiting its ability to function during overloads. Additionally, SIP, as a new area formed by land reclamation from lakes, has low-lying terrain and poor-draining artificial fill soil, weakening natural drainage. In contrast, GD relies

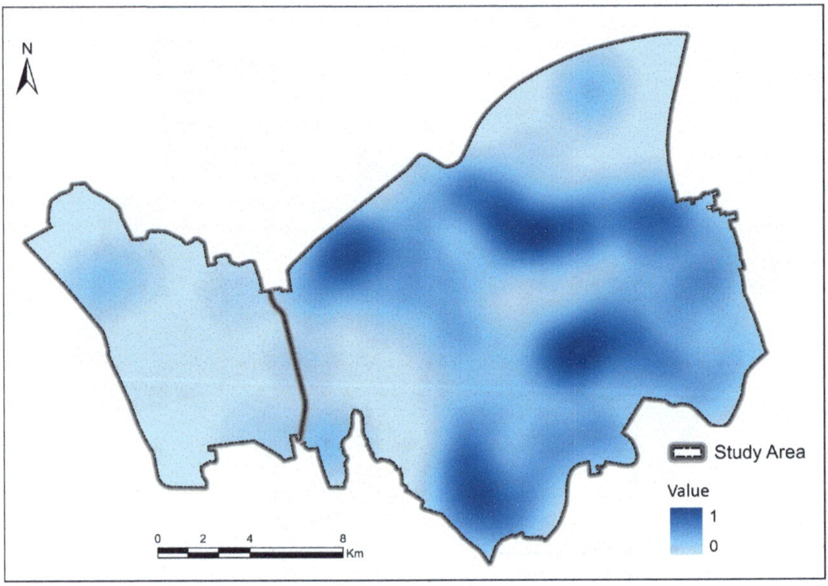

Fig. 4 Kernel density analysis of regional flooding sites (*Source* Made by authors)

on traditional rivers and a terrain-integrated natural drainage system, forming natural catchments that effectively regulate surface runoff.

4.2 Questionnaire Outcomes

A total of 151 valid questionnaires were collected for this study, including 70 from Gusu District and 81 from the Industrial Park District. The demographic profile of the participants is illustrated in Figs. 6 and 7.

Given the inclusion of Likert scale items in the questionnaire, it is reasonable to use Cronbach's alpha coefficient for reliability analysis to assess the internal consistency of the scale. In general, an alpha coefficient above 0.70 is considered to have an acceptable level of reliability (Cronbach 1951). The results of the reliability analysis in this study showed an alpha value of 0.840 (Table 4), which is a high level of consistency for the scale. In addition, Table 5 shows that the corrected item–total correlations for the two items were both 0.725, indicating a strong positive relationship between each item and the overall scale score. This suggests that the two items can reliably reflect respondents' perceptions of the distribution and impacts of GI in stormwater management.

To compare the current situations of GD and SIP in stormwater ponding issues, linear regression analyses were performed on three sets of variables by using SPSS. Firstly, the relationship between the study areas (1 = GD, 2 = SIP) and the frequency

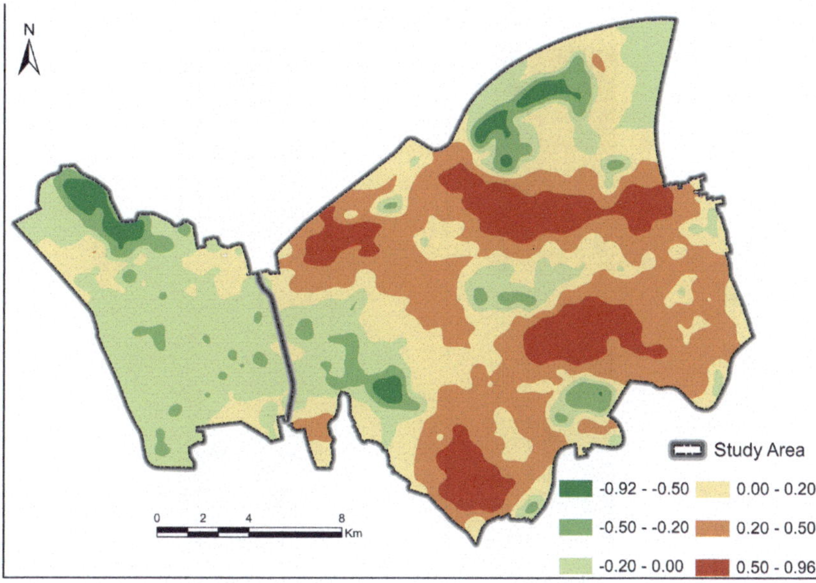

Fig. 5 Effectiveness of GI on flood mitigation (*Source* Made by authors)

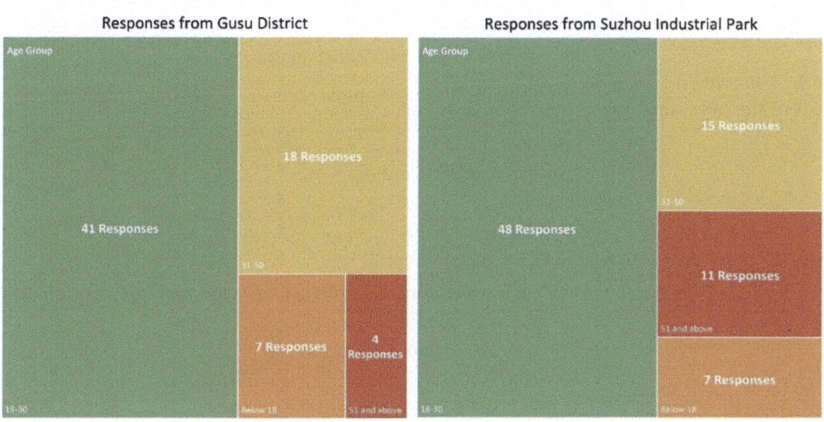

Fig. 6 The age group component of responders from GD and SIP (*Source* Made by authors)

of stormwater ponding events (1 = never, 5 = very frequently) was examined. The regression coefficient (B = 0.196, $p = 0.274$) indicates a non-significant positive relationship (Table 6). This result suggests that SIP experiences slightly more frequent stormwater ponding than GD, while the difference is not statistically significant as $p > 0.05$. Subsequently, this study assessed how the duration of stormwater recession (1 = <1 h, 6 = >24 h) differs between the two districts (Table 7). The regression coefficient (B = −0.081, $p = 0.936$) reveals a negligible and statistically non-significant

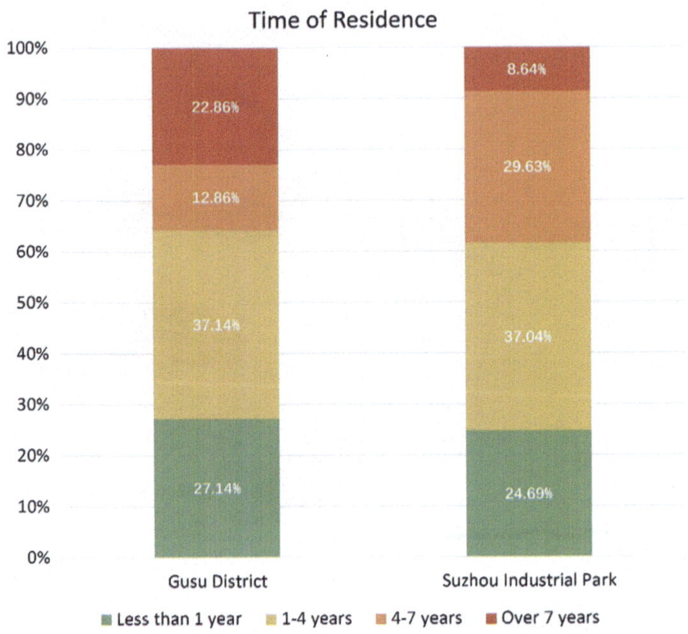

Fig. 7 The time of residence of responders from GD and SIP (*Source* Made by authors)

Table 4 The reliability analysis of the questionnaire

N of items	Sample size	Cronbach's alpha
2	151	0.840

Source Made by authors

Table 5 Item-total statistics table of the questionnaire

Items	Mean if item deleted	Variance if item deleted	Corrected item-total correlation
Q.12	3.30	1.307	0.725
Q.13	3.19	1.152	0.725

Source Made by authors

difference. The result implies that while SIP may exhibit marginally quicker water recession, this variation lacks statistical significance. Table 8 shows the statistical outcomes about the comparison of the extent to which residents in GD and SIP are affected by stormwater ponding issues. The coefficient ($B = 0.137, p = 0.514$) shows that residents in SIP reported slightly higher subjective ratings regarding the impact of stormwater issues. Nonetheless, the difference remains statistically insignificant.

In terms of evaluating the differences of the distribution and effectiveness of GI in the two districts, another three linear regression analyses were conducted in SPSS using the district as the independent variable. The dependent variables were:

Table 6 Coefficients table—district (Q2) and stormwater ponding frequency (Q4)

Predictor	B	Std. error	Beta	t	Sig. (p)
(Constant)	2.361	0.288	—	8.188	0.000
District (1 = GD, 2 = SIP)	0.196	0.179	0.090	1.098	0.274

Source Made by authors

Table 7 Coefficients table—district (Q2) and duration of stormwater recession (Q5)

Predictor	B	Std. error	Beta	t	Sig. (p)
(Constant)	2.671	0.412	–	6.477	0.000
District (1 = GD, 2 = SIP)	−0.021	0.255	−0.007	−0.081	0.936

Source Made by authors

Table 8 Coefficients table—district (Q2) and extent of stormwater ponding issue impacts (Q8)

Predictor	B	Std. error	Beta	t	Sig. (p)
(Constant)	2.561	0.339	—	7.559	0.000
District (1 = GD, 2 = SIP)	0.137	0.210	0.056	0.654	0.514

Source Made by authors

quantity of GI, distribution of GI, and effectiveness of GI in reducing stormwater ponding. All three models yielded statistically significant results, demonstrating the differences in GI applications between the two areas based on resident perceptions. From Table 9, it can be seen that B = 0.425 and $p = 0.029$, which reflects that SIP tends to score higher in terms of GI quantity than GD. This result may show that SIP covers more GI than GD. The next model examined the distribution situation of GI in GD and SIP (Table 10). The regression coefficient (B = 0.426, $p = 0.015$) indicates a significant positive relationship (Table 11), which suggests that SIP has a more even distribution of GI than GD. The last model also indicates the similar difference due to the regression coefficient (B = 0.408, $p = 0.028$), which shows that residents in SIP demonstrate greater recognition of the effectiveness of GI in addressing stormwater ponding issue.

Overall, the results of questionnaire indicate that while there is no statistically significant difference between GD and SIP in terms of the frequency, duration, and

Table 9 Coefficients table—district (Q2) and quantity of GI (Q11)

Predictor	B	Std. error	Beta	t	Sig. (p)
(Constant)	2.260	0.311	–	7.273	0.000
District (1 = GD, 2 = SIP)	0.425	0.192	0.178	2.211	0.029

Source Made by authors

Table 10 Coefficients table—district (Q2) and distribution of GI (Q12)

Predictor	B	Std. error	Beta	t	Sig. (*p*)
(Constant)	2.532	0.278	–	9.098	0.000
District (1 = GD, 2 = SIP)	0.426	0.172	0.198	2.471	0.015

Source Made by authors

Table 11 Coefficients table—district (Q2) and precepted effectiveness of GI in reducing stormwater ponding (Q13)

Predictor	B	Std. error	Beta	t	Sig. (*p*)
(Constant)	2.678	0.297	–	9.001	0.000
District (1 = GD, 2 = SIP)	0.408	0.184	0.179	2.216	0.028

Source Made by authors

perceived impact of stormwater ponding issues, there is a significant difference in residents' perceptions of GI. Specifically, SIP has a greater amount of green infrastructure and a more even spatial distribution compared to GD, which validated the GIS analyses results.

5 Discussion

A comparative analysis of green infrastructure (GI) between Gusu District (GD) and Suzhou Industrial Park (SIP) provides critical insights into the adaptability and effectiveness of GI under distinct urban development contexts. The findings reveal that GD exhibits higher GI coverage and a more fragmented landscape pattern dominated by small, scattered patches, which aligns with its micro-intervention greening strategy. This strategy emphasizes incremental upgrades of pocket green spaces within the constrained urban fabric, effectively enhancing local ecological functions and providing accessible green spaces for residents. Subjective perceptions from residents further indicate that GD's GI distribution is relatively abundant and even, with significant perceptual differences in GI quantity and spatial configuration between GD and SIP. These differences validate GD's success in improving stormwater management and overall environmental quality. However, residents in GD still report rainwater ponding in pedestrian streets and underground parking lots or low-lying entrances (Table 12), likely due to the low permeability of traditional paving materials (e.g., stone slabs) and heritage preservation constraints that limit drainage retrofits.

In contrast, SIP demonstrates lower GI coverage and higher impervious surface ratios, reflecting its intensive urbanization and associated challenges in surface runoff management. Despite SIP's larger aggregate GI area, spatial mismatches between

Table 12 Rainwater ponding locations

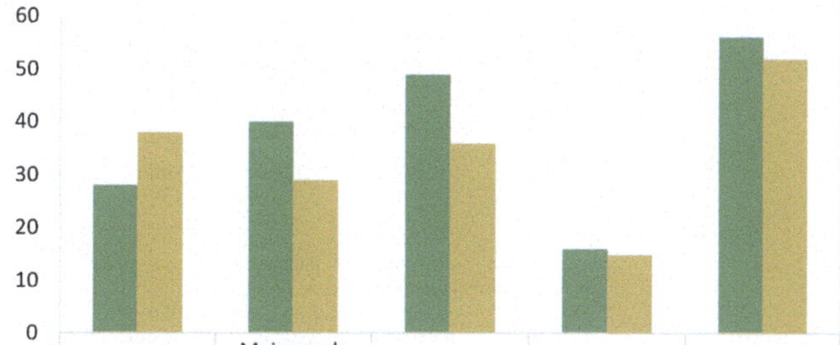

	Internal roads of the communities	Main roads and near public transportation stations	Pedestrian streets	Parks or green spaces	Underground parking lots or low-lying entrances
Gusu District	28	40	49	16	56
Suzhou Industrial Park	38	29	36	15	52

Source Made by authors

concentrated impervious zones and GI distribution patterns likely constrain its hydrological regulation capacity. Kernel density analysis highlights localized deficiencies in runoff mitigation efficacy. While SIP's planned ecological features—such as wetland parks and green corridors—contribute to its overarching ecological framework, their hydrological benefits are diminished in densely paved industrial and residential areas. Questionnaire results corroborate this limitation, with SIP residents identifying *internal roads of neighborhoods* and *underground parking lots or low-lying entrances* as rainwater ponding hotspots (Table 12), underscoring the inadequacy of centralized GI planning in addressing localized runoff accumulation.

GIS analysis further demonstrates GD's superior effectiveness in flood risk mitigation, as evidenced by predominantly negative values in runoff density-GI density differential analysis. This is attributed to GD's reliance on traditional canal networks and terrain-integrated natural drainage systems, which form natural catchments to regulate surface runoff. Conversely, SIP's low-lying topography and poorly permeable artificial fill soils weaken natural drainage capacity. These limitations, combined with potential undercapacity of GI designs to account for extreme rainfall intensity under climate change, restrict SIP's operational resilience during overload scenarios.

Resident surveys on GI improvement priorities highlight cross-district consensus: both areas advocate for enhanced runoff control through measures such as increasing permeable pavement coverage, expanding green spaces, and implementing green roofs. These findings underscore the urgent need to align GI planning with localized hydrological challenges and climate resilience goals (Table 13).

Table 13 Desired improvements to district's GI

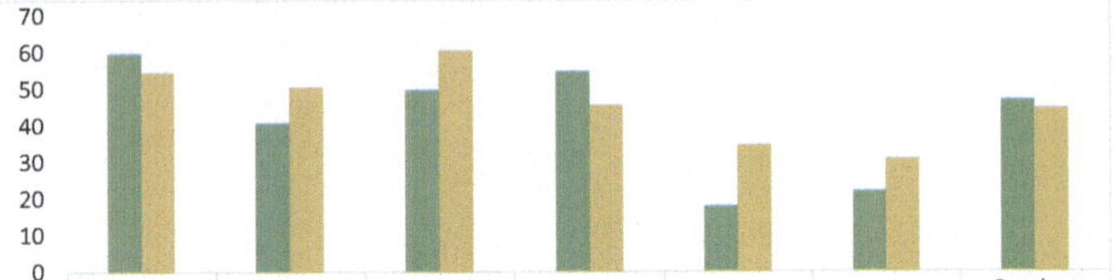

	Increase the coverage of permeable pavements	Add more rain gardens, vegetation strips, and urban green belts	Create more public green spaces, community gardens, and green areas	Construct green roofs	Strengthen rainwater harvesting systems and improve drainage facilities	Protect existing natural areas, wetlands, and ecological corridors	Conduct regular maintenance and upgrades of existing green infrastructure
Gusu District	60	41	50	55	18	22	47
Suzhou Industrial Park	55	51	61	46	35	31	45

Source Made by authors

6 Conclusion

This study systematically examines the adaptation and effectiveness of GI in two different urban environments, GD and SIP, to advance climate-resilient stormwater management strategies. GD's GI prioritizes decentralized, small-scale interventions, while SIP relies on engineered wetlands and green corridors. GD's decentralized but fully functional GI network outperforms SIP's centralized GI planning in mitigating flood risk under extreme rainfall scenarios. Despite the larger total area of GI in SIP, its hydrological efficacy is constrained by the concentration of impervious surfaces and artificial topographic constraints.

These findings offer recommendations for sponge cities in China. For historic old towns, GI and blue infrastructure can work together to influence surface runoff by increasing the coverage of permeable pavement without compromising cultural authenticity. As for the construction of new districts, the spatial mismatch between centralized impermeable zones and Gl distribution needs to be addressed. Permeable pavements and vegetated depressions are rationalized to enhance infiltration and reduce peak flows. Meanwhile, the SIP can learn from GD's experience and make similar progressive greening improvements in its area. Furthermore, design natural drainage systems that are compatible with the topography to improve the efficiency of stormwater management.

In terms of limitations, several factors may have influenced the findings. The land cover dataset may inadequately represent small-scale GI elements, particularly in GD's dense historical urban fabric. For instance, pocket green spaces or narrow vegetated alleys might be misclassified or omitted, potentially underestimating GD's GI connectivity and functional diversity. Similarly, SIP's engineered wetlands and green roofs, which require high-resolution spatial delineation, might not be fully captured, affecting hydrological modeling precision. In addition, the precipitation data used in this study is based on the 10-year maximum daily precipitation extremes to model flooding risk. While this approach is effective in identifying vulnerable areas under extreme events, it may ignore differences in hydrologic response under normalized rainfall scenarios.

Limitations in the questionnaire, first, although the total number of valid responses was 151, there was an imbalance between the two districts (70 from GD and 81 from SIP), which may have introduced bias and affected the comparative reliability of the statistical outcomes. Second, during extreme rainfall events, residents tend to reduce outdoor activities, limiting their opportunities to observe and report on stormwater ponding. It could be a factor that can explain why GD and SIP perform similarly in stormwater ponding issues. Additionally, while the study relied on respondents' self-reported perceptions and observations, such data may be subject to personal biases and inaccuracies, especially concerning the identification and understanding of various GI types despite the survey's visual aids. Finally, the study only captured a snapshot of resident experiences during a single season and may not reflect variations across longer temporal scales or under different climatic conditions.

Given these insights, future research should consider integrating real-time and multi-source data, such as social media, remote sensing, hydrological sensors, and rainfall diaries, to provide a more objective and comprehensive understanding of localized flooding phenomena. It is also recommended to increase sample sizes and ensure regional balance to enhance representativeness and explanatory power. Participatory mapping and in-depth interviews could further help unpack how everyday lived experiences shape public perceptions of GI effectiveness, offering targeted insights for urban planning, public engagement, and sustainable GI implementation strategies.

Acknowledgements All the authors of this chapter would like to express the gratitude to the book editors, especially Prof. Ali Cheshmehzangi, for the efforts in the publication of this book. In addition, we would like to thank all the respondents of questionnaires in GD and SIP in Suzhou.

References

Addo-Bankas O, Wei T, Zhao Y, Bai X, Núñez AE, Stefanakis A (2024) Revisiting the concept, urban practices, current advances, and future prospects of green infrastructure. Sci Total Environ 954. Elsevier B.V. https://doi.org/10.1016/j.scitotenv.2024.176473

Ahern J (2007) Green infrastructure for cities: the spatial dimension

Ayoubi Ayoublu S, Vafakhah M, Pourghasemi HR (2024) Efficiency evaluation of low impact development practices on urban flood risk. J Environ Manage 356. https://doi.org/10.1016/j.jenvman.2024.120467

Bartesaghi Koc C, Osmond P, Peters A (2016) A green infrastructure typology matrix to support urban microclimate studies. Procedia Eng 169:183–190. https://doi.org/10.1016/j.proeng.2016.10.022

Bartesaghi Koc C, Osmond P, Peters A (2017) Towards a comprehensive green infrastructure typology: a systematic review of approaches, methods and typologies. Urban Ecosyst 20(1):15–35. https://doi.org/10.1007/s11252-016-0578-5

Beecham S, Razzaghmanesh M, Bustami R, Ward J (2018) The role of green roofs and living walls as WSUD approaches in a dry climate. In: Approaches to water sensitive urban design: potential, design, ecological health, urban greening, economics, policies, and community perceptions. Elsevier, pp 409–430. https://doi.org/10.1016/B978-0-12-812843-5.00020-4

Berland A, Shiflett SA, Shuster WD, Garmestani AS, Goddard HC, Herrmann DL, Hopton ME (2017) The role of trees in urban stormwater management. Landsc Urban Planning 162:167–177. Elsevier B.V. https://doi.org/10.1016/j.landurbplan.2017.02.017

Bonciarelli L, Orlandi F, Muscas D, Fornaciari M (2025) Sustainable stormwater management and bioretention: an overview of reviews of the last 10 years. Land 14(4). Multidisciplinary Digital Publishing Institute (MDPI). https://doi.org/10.3390/land14040736

Booth DB, Leavitt J (1999) Field evaluation of permeable pavement systems for improved stormwater management. J Am Plann Assoc 65(3):314–325. https://doi.org/10.1080/01944369908976060

Bullock A, Acreman M (2003) The role of wetlands in the hydrological cycle. Hydrol Earth Syst Sci 7(3)

Chan FKS, Griffiths JA, Higgitt D, Xu S, Zhu F, Tang YT, Xu Y, Thorne CR (2018) "Sponge City" in China—a breakthrough of planning and flood risk management in the urban context. Land Use Policy 76:772–778. https://doi.org/10.1016/j.landusepol.2018.03.005

Chen W, Wang W, Huang G, Wang Z, Lai C, Yang Z (2021) The capacity of grey infrastructure in urban flood management: a comprehensive analysis of grey infrastructure and the green-grey approach. Int J Disaster Risk Reduct 54. https://doi.org/10.1016/j.ijdrr.2021.102045

Chini CM, Canning JF, Schreiber KL, Peschel JM, Stillwell AS (2017) The green experiment: cities, green stormwater infrastructure, and sustainability. Sustainability (Switzerland) 9(1). https://doi.org/10.3390/su9010105

Cronbach LJ (1951) Coefficient alpha and the internal structure of tests. Psychometrika 16(3)

Davidson C (1998) Issues in measuring landscape fragmentation. Wildlife Soc Bull 32–37

Denman EC, May PB, Moore GM (2011) The use of trees in urban stormwater management

Dong X, Guo H, Zeng S (2017) Enhancing future resilience in urban drainage system: green versus grey infrastructure. Water Res 124:280–289. https://doi.org/10.1016/j.watres.2017.07.038

Dowtin AL, Cregg BC, Nowak DJ, Levia DF (2023) Towards optimized runoff reduction by urban tree cover: a review of key physical tree traits, site conditions, and management strategies. Landsc Urban Planning 239. Elsevier B.V. https://doi.org/10.1016/j.landurbplan.2023.104849

Fan C, Myint S (2014) A comparison of spatial autocorrelation indices and landscape metrics in measuring urban landscape fragmentation. Landsc Urban Plan 121:117–128

Ferrans P, Torres MN, Temprano J, Rodríguez Sánchez JP (2022) Sustainable urban drainage system (SUDS) modeling supporting decision-making: a systematic quantitative review. Sci Total Environ 806. Elsevier B.V. https://doi.org/10.1016/j.scitotenv.2021.150447

Fletcher TD, Shuster W, Hunt WF, Ashley R, Butler D, Arthur S, Trowsdale S, Barraud S, Semadeni-Davies A, Bertrand-Krajewski JL, Mikkelsen PS, Rivard G, Uhl M, Dagenais D, Viklander M (2015) SUDS, LID, BMPs, WSUD and more – the evolution and application of terminology surrounding urban drainage. Urban Water J 12(7):525–542. https://doi.org/10.1080/1573062X.2014.916314

Grabowski ZJ, McPhearson T, Matsler AM, Groffman P, Pickett STA (2022) What is green infrastructure? A study of definitions in US city planning. Front Ecol Environ 20(3):152–160. Wiley. https://doi.org/10.1002/fee.2445

Gusu District People's Government. (2024). Introduction to Gusu District [in Chinese]. https://www.gusu.gov.cn/gsq/zjgs/zjgs.shtml

Hanna E, Bruno D, Comín FA (2024) The ecosystem services supplied by urban green infrastructure depend on their naturalness, functionality and imperviousness. Urban Ecosyst 27(1):187–202. https://doi.org/10.1007/s11252-023-01442-9

Hansen R, Pauleit S (2014) From multifunctionality to multiple ecosystem services? A conceptual framework for multifunctionality in green infrastructure planning for urban areas. Ambio 43(4):516–529. https://doi.org/10.1007/s13280-014-0510-2

Hu JL, Miao CY (2023) CHM_PRE V2: an upgraded high-precision gridded precipitation dataset for the Chinese mainland considering spatial autocorrelation and covariates. National Tibetan Plateau / Third Pole Environment Data Center. https://doi.org/10.11888/Atmos.tpdc.300523 (2023).

Ignatieva M, Mofrad F (2023) Understanding urban green spaces typology's contribution to comprehensive green infrastructure planning: a study of Canberra, the National Capital of Australia. Land 12(5). https://doi.org/10.3390/land12050950

Ishimatsu K, Ito K, Mitani Y, Tanaka Y, Sugahara T, Naka Y (2017) Use of rain gardens for stormwater management in urban design and planning. Landsc Ecol Eng 13(1):205–212. https://doi.org/10.1007/s11355-016-0309-3

Jiang Y, Zevenbergen C, Ma Y (2018) Urban pluvial flooding and stormwater management: a contemporary review of China's challenges and "sponge cities" strategy. Environ Sci Policy 80:132–143. Elsevier Ltd. https://doi.org/10.1016/j.envsci.2017.11.016

Jones L, Anderson S, Læssøe J, Banzhaf E, Jensen A, Bird DN, Miller J, Hutchins MG, Yang J, Garrett J, Taylor T, Wheeler BW, Lovell R, Fletcher D, Qu Y, Vieno M, Zandersen M (2022) A typology for urban green infrastructure to guide multifunctional planning of nature-based solutions. Nat-Based Solutions 2:100041. https://doi.org/10.1016/j.nbsj.2022.100041

Lashford C, Rubinato M, Cai Y, Hou J, Abolfathi S, Coupe S, Charlesworth S, Tait S (2019) SuDS & sponge cities: a comparative analysis of the implementation of pluvial flood management in the UK and China. Sustainability (Switzerland) 11(1). MDPI. https://doi.org/10.3390/su11010213

Lee YC, Liu YF (2023) Co-benefits of preserving urban farmland as climate change adaptation strategy: an emergy approach. Ecol Ind 154. https://doi.org/10.1016/j.ecolind.2023.110722

Lei Y (2024) UoNJC and ANR: the two UK case studies of stormwater management through SuDS. In Cheshmehzangi A, Sedrez M, Flynn A (ed) Rethinking stormwater management through sustainable urban design. Springer, Singapore, pp 89–115. https://doi.org/10.1007/978-981-97-4924-9_5

Li X, Li J, Fang X, Gong Y, Wang W (2016) Case studies of the sponge city program in China. World environmental and water resources congress 2016: watershed management, irrigation and drainage, and water resources planning and management - papers from sessions of the proceedings of the 2016 world environmental and water resources congress, pp 295–308. https://doi.org/10.1061/9780784479858.031

Liu H, Jia Y, Niu C (2017) "Sponge city" concept helps solve China's urban water problems. Environ Earth Sci 76(14). https://doi.org/10.1007/s12665-017-6652-3

Liu T, Lawluvy Y, Shi Y, Yap PS (2021) Low impact development (LID) practices: a review on recent developments, challenges and prospects. Water, Air, Soil Pollut 232(9). Springer Science and Business Media Deutschland GmbH. https://doi.org/10.1007/s11270-021-05262-5

Li ZH, He W, Cheng MF, Hu JX, An X, Huang Y, Yang GY, Zhang HY (2023) (2023) SinoLC-1: the first 1-meter resolution national-scale land-cover map of China created with a deep learning framework and open-access data. Zenodo, 2.4. https://doi.org/10.5281/zenodo.8214871

Malaviya P, Sharma R, Sharma PK (2019) Rain gardens as stormwater management tool. Sustain Green Technol Environ Manage 141–166. Springer Singapore. https://doi.org/10.1007/978-981-13-2772-8_7

Matsler AM, Meerow S, Mell IC, Pavao-Zuckerman MA (2021) A 'green' chameleon: exploring the many disciplinary definitions, goals, and forms of "green infrastructure." Landsc Urban Planning 214. Elsevier B.V. https://doi.org/10.1016/j.landurbplan.2021.104145

Mell IC (2008) Green infrastructure: concepts and planning. FORUM ejournal 8(1):69–80. Newcastle University, Newcastle, UK

Milazzo F, Francksen RM, Zavattaro L, Abdalla M, Hejduk S, Enri SR, Pittarello M, Price PN, Schils RLM, Smith P, Vanwalleghem T (2023) The role of grassland for erosion and flood mitigation in Europe: a meta-analysis. Agric, Ecosyst Environ 348. Elsevier B.V. https://doi.org/10.1016/j.agee.2023.108443

Monteiro R, Ferreira JC, Antunes P (2020) Green infrastructure planning principles: an integrated literature review. Land 9(12):1–19. https://doi.org/10.3390/land9120525

Nazarpour S, Gnecco I, Palla A (2023) Evaluating the effectiveness of bioretention cells for urban stormwater management: a systematic review. Water (Switzerland) 15(5). MDPI. https://doi.org/10.3390/w15050913

Nguyen TT, Ngo HH, Guo W, Wang XC, Ren N, Li G, Ding J, Liang H (2019) Implementation of a specific urban water management - Sponge City. Sci Total Environ 652:147–162. Elsevier B.V. https://doi.org/10.1016/j.scitotenv.2018.10.168

Nieuwenhuijsen MJ (2025) Green infrastructure and health. Guest 42:29. www.Annualreviews.Org.Guest. https://doi.org/10.1146/annurev-publhealth

Power AG (2010) Ecosystem services and agriculture: tradeoffs and synergies. Philos Trans Roy Soc B: Biol Sci 365(1554):2959–2971. Royal Society. https://doi.org/10.1098/rstb.2010.0143

Raimondi A, Becciu G (2021) Performance of green roofs for rainwater control. Water Resour Manage 35(1):99–111. https://doi.org/10.1007/s11269-020-02712-3

Razzaghmanesh M, Beecham S, Kazemi F (2014) Impact of green roofs on stormwater quality in a South Australian urban environment. Sci Total Environ 470–471:651–659. https://doi.org/10.1016/j.scitotenv.2013.10.047

Rodak CM, Moore TL, David R, Jayakaran AD, Vogel JR (2019) Urban stormwater characterization, control, and treatment. Water Environ Res 91(10):1034–1060. https://doi.org/10.1002/wer.1173

Rooney RC, Foote L, Krogman N, Pattison JK, Wilson MJ, Bayley SE (2015) Replacing natural wetlands with stormwater management facilities: biophysical and perceived social values. Water Res 73:17–28. https://doi.org/10.1016/j.watres.2014.12.035

Seitz J, Escobedo F (2008) Urban forests in Florida: trees control stormwater runoff and improve water quality. http://edis.ifas.ufl.edu

Seiwert A, Rößler S (2020) Understanding the term green infrastructure: origins, rationales, semantic content and purposes as well as its relevance for application in spatial planning. Land Use Policy, 97. https://doi.org/10.1016/j.landusepol.2020.104785

Selbig WR, Loheide SP, Shuster W, Scharenbroch BC, Coville RC, Kruegler J, Avery W, Haefner R, Nowak D (2022) Quantifying the stormwater runoff volume reduction benefits of urban street tree canopy. Sci Total Environ 806. https://doi.org/10.1016/j.scitotenv.2021.151296

Semeraro T, Scarano A, Buccolieri R, Santino A, Aarrevaara E (2021) Planning of urban green spaces: an ecological perspective on human benefits. Land 10(2):1–26. MDPI AG. https://doi.org/10.3390/land10020105

Seyedashraf O, Bottacin-Busolin A, Harou JJ (2022) A design framework for considering spatial equity in sustainable urban drainage infrastructure. Sustain Cities Soc 85. https://doi.org/10.1016/j.scs.2022.103960

Shafique M, Kim R, Kyung-Ho K (2018) Green roof for stormwater management in a highly urbanized area: the case of Seoul, Korea. Sustainability (Switzerland) 10(3). https://doi.org/10.3390/su10030584

Sharma R, Malaviya P (2021) Management of stormwater pollution using green infrastructure: the role of rain gardens. Wiley Interdisc Rev: Water 8(2). Wiley. https://doi.org/10.1002/wat2.1507

Stovin V (2010) The potential of green roofs to manage urban stormwater. Water and Environ J 24(3):192–199. https://doi.org/10.1111/j.1747-6593.2009.00174.x

Suppakittpaisarn P, Jiang X, Sullivan WC (2017) Green infrastructure, green stormwater infrastructure, and human health: a review. Curr Landsc Ecol Rep 2(4):96–110. https://doi.org/10.1007/s40823-017-0028-y

Suzhou Municipal People's Government (2024a) Natural geography and resources [in Chinese]. https://www.suzhou.gov.cn/szsrmzf/szsqsl2024/202406/09db8be4137c4f94b06a3b8136d23a53.shtml

Suzhou Municipal People's Government (2024b) Suzhou's permanent population reaches 12.958 million, remaining the largest in Jiangsu Province, with an increase of 47,400 compared to the end of 2022 [in Chinese]. https://www.suzhou.gov.cn/szsrmzf/szyw/202403/815da32393f143008a2806ad3ffe6618.shtml

Tang S, Luo W, Jia Z, Liu W, Li S, Wu Y (2016) Evaluating retention capacity of infiltration rain gardens and their potential effect on urban stormwater management in the sub-humid loess

region of China. Water Resour Manage 30(3):983–1000. https://doi.org/10.1007/s11269-015-1206-5
Tian Y, Jim CY, Tao Y, Shi T (2011) Landscape ecological assessment of green space fragmentation in Hong Kong. Urban Forest Urban Greening 10(2):79–86
Trowsdale SA, Simcock R (2011) Urban stormwater treatment using bioretention. J Hydrol 397(3–4):167–174. https://doi.org/10.1016/j.jhydrol.2010.11.023
Tzoulas K, Korpela K, Venn S, Yli-Pelkonen V, Kaźmierczak A, Niemela J, James P (2007) Promoting ecosystem and human health in urban areas using green infrastructure: a literature review. Landsc Urban Planning 81(3):167–178. Elsevier. https://doi.org/10.1016/j.landurbplan.2007.02.001
Vijayaraghavan K, Biswal BK, Adam MG, Soh SH, Tsen-Tieng DL, Davis AP, Chew SH, Tan PY, Babovic V, Balasubramanian R (2021) Bioretention systems for stormwater management: recent advances and future prospects. J Environ Manage 292. Academic Press. https://doi.org/10.1016/j.jenvman.2021.112766
Vineyard D, Ingwersen WW, Hawkins TR, Xue X, Demeke B, Shuster W (2015) Comparing green and grey infrastructure using life cycle cost and environmental impact: a rain garden case study in Cincinnati, OH. J Am Water Resour Assoc 51(5):1342–1360. https://doi.org/10.1111/1752-1688.12320
Vogel JR, Moore TL (2016) Urban stormwater characterization, control, and treatment. Water Environ Res 88(10):1918–1950. https://doi.org/10.2175/106143016x14696400495938
Vymazal J (2007) Removal of nutrients in various types of constructed wetlands. Sci Total Environ 380(1–3):48–65. https://doi.org/10.1016/j.scitotenv.2006.09.014
Wang H, Mei C, Liu JH, Shao WW (2018) A new strategy for integrated urban water management in China: Sponge city. Sci China Technol Sci 61(3):317–329. Springer. https://doi.org/10.1007/s11431-017-9170-5
Wang J, Banzhaf E (2018) Towards a better understanding of green infrastructure: a critical review. Ecol Ind 85:758–772. Elsevier B.V. https://doi.org/10.1016/j.ecolind.2017.09.018
Wong THF, Breen PF, Somes NLG (1999) Ponds vs wetlands – performance considerations in stormwater quality management
Xia J, Zhang YY, Xiong LH, He S, Wang LF, Yu ZB (2017) Opportunities and challenges of the Sponge City construction related to urban water issues in China. Sci China Earth Sci 60(4):652–658. https://doi.org/10.1007/s11430-016-0111-8
Xie J, Wu C, Li H, Chen G (2017) Study on storm-water management of grassed swales and permeable pavement based on SWMM. Water (Switzerland) 9(11). https://doi.org/10.3390/w9110840
Xing YJ, Chen TL, Gao MY, Pei SL, Pan WB, Chiang PC (2021) Comprehensive performance evaluation of green infrastructure practices for urban watersheds using an engineering–environmental–economic (3E) model. Sustainability (Switzerland) 13(9). https://doi.org/10.3390/su13094678
Xu C, Tang T, Jia H, Xu M, Xu T, Liu Z, Long Y, Zhang R (2019) Benefits of coupled green and grey infrastructure systems: evidence based on analytic hierarchy process and life cycle costing. Resour, Conserv Recycl 151. https://doi.org/10.1016/j.resconrec.2019.104478
Yang Y, Chui TFM (2018) Optimizing surface and contributing areas of bioretention cells for stormwater runoff quality and quantity management. J Environ Manage 206:1090–1103. https://doi.org/10.1016/j.jenvman.2017.11.064
Yang Y, Li Z, Li P, Ren Z, Gao H, Wang T, Xu G, Yu K, Shi P, Tang S (2017) Variations in runoff and sediment in watersheds in loess regions with different geomorphologies and their response to landscape patterns. Environ Earth Sci 76:1–11
Ying J, Zhang X, Zhang Y, Bilan S (2022) Green infrastructure: systematic literature review. Econ Res-Ekonomska Istrazivanja 351:343–366. Taylor and Francis Ltd. https://doi.org/10.1080/1331677X.2021.1893202
Zedler JB (2003) Wetlands at your service: reducing impacts of agriculture at the watershed scale. www.frontiersinecology.org

Zhang L, Ye Z, Shibata S (2020) Assessment of rain garden effects for the management of urban storm runoff in Japan. Sustainability (Switzerland) 12(23):1–17. https://doi.org/10.3390/su12239982
Zhang Z, Miao Y, Li J, Liu D, Fang X (2019) Improvement effect of rainfall source control facilities on urban drainage capacity in different regions of China. J Hydrol 579. https://doi.org/10.1016/j.jhydrol.2019.124127

The Microclimate and Human Thermal Comfort Impacts of Increased Vegetation Complexity in Urban Streets and Greenspaces

Mohammad A. Rahman, Pui Kwan Cheung, Nayanesh Pattnaik, Theresa W. L. Lam, Marie C. Dade, C. Y. Jim, Stephan Pauleit, and Stephen J. Livesley

Abstract There is increasing emphasis on enhancing vegetation complexity in urban landscapes to deliver both ecological and social benefits, including improved human wellbeing, nature connectedness, and biodiversity support. Vegetation complexity achieved through combinations of trees, shrubs and grasses can be introduced in both urban greenspaces and streetscapes. While it is often assumed that such complexity improves urban microclimates and human thermal comfort, these effects are highly context-dependent, shaped by background climate, urban morphology, vegetation spatial structure, and plant traits. To investigate these dynamics, this chapter integrates a critical review of the literature with empirical data from three case study cities representing contrasting climate zones: Melbourne and Munich (temperate) and Hong Kong (humid subtropical). Using comparable methods, we assess the influence of multi-layered vegetation on human thermal comfort through in-situ microclimate measurements, applying standardized indices such as physiological equivalent temperature (PET) and universal thermal climate index (UTCI). Positive impacts are generally driven by enhanced shading and evapotranspiration, while potential drawbacks—such as reduced wind flow, increased humidity, and hindered pollutant dispersion may moderate benefits, especially in dense urban settings. Our empirical findings show that multi-layered vegetation consistently reduces thermal stress in greenspaces. In streetscapes; however, the outcomes are more variable due to constrained airflow, vegetation configuration, and local built form. At the same

M. A. Rahman (✉) · P. K. Cheung · T. W. L. Lam · M. C. Dade · S. J. Livesley
School of Agriculture, Food and Ecosystem Sciences, The University of Melbourne, Burnley, VIC, Australia
e-mail: mohammad.a.rahman@unimelb.edu.au

N. Pattnaik · S. Pauleit
Strategic Landscape Planning and Management, School of Life Sciences, Technical University of Munich, Munich, Germany

C. Y. Jim
Department of Social Sciences and Policy Studies (SSPS), The Education University of Hong Kong, Hong Kong, China

A. Cheshmehzangi et al. (eds.), *Nature in Cities, Nurturing Cities*, Urban Sustainability,
https://doi.org/10.1007/978-981-95-5487-4_6

time, literature review findings using modeled output often diverged from empirical findings, emphasizing the need for context-specific calibration and ecological understanding. This study supports a shift toward integrated design thinking, where vegetation complexity is purposefully calibrated to balance thermal comfort, biodiversity, and urban function. Future cooling strategies must consider not only vegetation presence but also its growth-form combinations and interaction with built environments. To inform this, long-term, field-based studies on fully complex vegetation structures are needed to capture diurnal and seasonal variability in diverse urban contexts.

Keywords Microclimate · Urban heat · Green infrastructure · Vegetation structure · Human thermal comfort

1 Introduction

Urban nature-based solutions that aim to increase nature within cities often involve increasing the structural complexity of vegetation within urban greenspaces and streets, as vegetation complexity is a well-regarded approach for improving biodiversity habitat and nature-connectedness (Garrard et al. 2018; Harris et al. 2018; Tan et al. 2022; Threlfall et al. 2017). Vegetation complexity encompasses the diversity and spatial arrangement of vegetation elements across both vertical and horizontal dimensions within urban environments. However, increased vegetation complexity in urban greenspaces and streets will also influence the urban microclimate, by changing solar radiation inputs, air movement and humidity in those spaces (Lai et al. 2019). Understanding how the structural complexity of vegetation influences local microclimate will help ensure that nature-based solutions in cities are implemented that both improve the socio-ecological benefits provided by nature and local microclimates for people and nature.

Mitigating urban heat is a priority for many cities. For society, urban heat can lead to reduced human comfort and risks of heat-related illnesses, hospitalization and mortality, as well as higher energy consumption for cooling, and exacerbated air pollution impacts (Oke et al. 2017). For wildlife, urban heat can lead to a decrease in species diversity, reproduction success, greater foraging requirements and under extreme events heatstroke and mass mortality (McDonald et al. 2008; Nazish et al. 2024). However, reducing urban heat is challenging as urban expansion and urban densification with limited green-and-blue infrastructures significantly contribute to increasing urban heat, and this impact is further compounded by global warming, urban heat island effect, and extreme heatwave events (Dodman et al. 2022). For example, Huang et al. (2019) projected increases in average summer daytime and nighttime air temperatures of 0.5–0.7 °C, and up to ~3 °C by 2050 in many cities, which can be up to twice the increase expected from greenhouse gas-induced climate change alone.

Vegetation is a key nature-based tool used by cities to mitigate urban heat and improve human thermal comfort (Esperón-Rodríguez et al. 2025; Iungman et al.

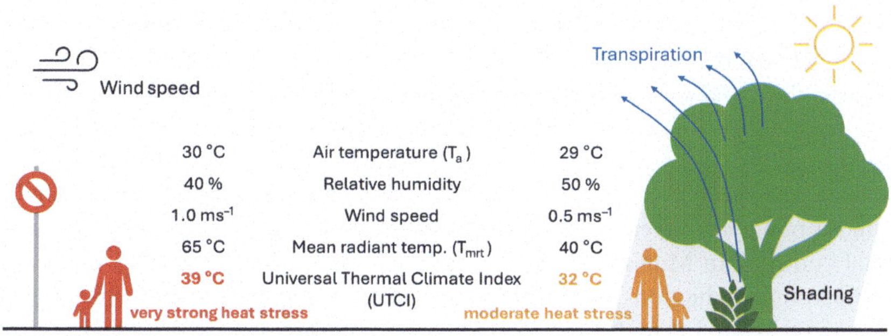

Fig. 1 Conceptual diagram indicating possible differences in four local microclimate variables (air temperature, relative humidity, wind speed and T_{mrt}) between an open urban setting (LHS) and under tree canopy (RHS). Human thermal comfort, as indicated by UTCI, is estimated from these four variables and demonstrates the capacity of urban vegetation to reduce human thermal stress from very strong to moderate. The human thermal benefit of vegetation is mainly driven by shading also with higher reflection compared to the built surfaces, and transpiration which decreases air temperature and T_{mrt} and is only partially offset by a small increase in relative humidity and decrease in wind speed

2023; Norton et al. 2015). Human thermal comfort is defined as "a condition of mind, which expresses satisfaction with the surrounding environment" (Abdel-Ghany et al. 2013). Human thermal comfort is subjectively assessed according to human clothing and activity categories, but more importantly from the measurement of four local microclimate variables (Fig. 1): air temperature, relative humidity, mean radiant temperature (T_{mrt}) and wind speed (Fanger 1970). Recent advances in the estimation of human thermal comfort in outdoor environments for biometeorology assessments has led to the development of the Universal Thermal Climate Index (UTCI) (Bröde et al. 2012), among a large range of other indices (Staiger et al. 2019).

Urban vegetation impacts the local microclimate and the four variables used to estimate human thermal comfort through three key processes (Fig. 1). Firstly, vegetation reflects and absorbs solar radiation which reduces the amount of solar radiation that reaches the ground or impervious surfaces (e.g., asphalt, concrete, or roofs). This reduction in solar radiation is what we perceive as shade cast, and includes shade cast by tree crowns, but also the shade cast by turfgrass on the soil below. The critical difference being that human bodies are able to directly benefit from the shade cast by overhead vegetation such as tree crowns. As less solar radiation reaches the permeable or impervious ground surfaces, less solar radiation is converted to sensible heat and accumulated in the urban landscape, which in turn lessens air temperature and mean radiant temperature (T_{mrt}) increase and improves human thermal comfort. Previous studies have shown the substantial reductions in surface and air temperature provided by vegetation shading (Armson et al. 2012; Pattnaik et al. 2024; Rahman et al. 2019). Tree canopies can reduce the input of shortwave radiation to the ground level by about 60–90% leading to reductions in surface temperature up to 40 °C (Rahman et al. 2020a).

Secondly, vegetation can alter the local microclimate in cities through transpiration. Transpiration is the process by which liquid water is converted to water vapor in a leaf and this vapor is then released into the atmosphere (Rahman et al. 2020b). This conversion of liquid water to water vapor makes use of the heat energy accumulated in the leaves from absorbing solar radiation, and this transpiration process prevents leaves from over-heating or wilting. Transpiration can effectively cool the leaf and a boundary layer of air around the vegetation which can improve human thermal comfort. In Munich, Germany, Rahman et al. (2018) reported the average air temperature reductions within the tree canopy compared to the adjacent open areas ranged from 1.8 to 1.3 °C for the dense canopy of *Tilia cordata*, and from 1.5 to 0.5 °C for the lighter canopy of *Robinia pseudoacacia*, with cooling effects diminishing as ambient temperatures increased. At the same time, air temperature reduction at pedestrian height (1.3 m above ground) was significantly reduced with the ready admixture of warmer air from the surrounding environment. However, transpiration can also negatively impact human thermal comfort by increasing the relative humidity of the local microclimate. High relative humidity can reduce the efficiency of sweat evaporation, and therefore evaporative cooling, from a human body (Ma et al. 2018).

Thirdly, urban vegetation can impact the local microclimate by changing wind speeds, boundary layer heights and turbulence. Some vegetation, such as street trees, hedges, or treed greenspace can significantly reduce wind speeds which reduces the convection cooling benefit to humans on warm summer days and can reduce the efficiency of sweat evaporative cooling as moist air is not replaced with dry as rapidly (Oke 1989). Furthermore, reduced wind speeds from vegetation can reduce the dispersion of water vapor released through transpiration thereby increasing the relative humidity of the local microclimate. Similarly, reduced wind speeds due to large vegetation elements can prevent the dissipation of warm air (especially at night) and prevent the dispersion of air pollutants emitted by vehicles from within a streetscape thereby enhancing the exposure of pedestrians to air pollution. However, in cool winter months, reduced wind speed from vegetation in urban landscapes can be positive for human thermal comfort, as it reduces body heat loss (Xu et al. 2018).

These three key processes by which urban vegetation impacts local microclimates, and thereby positively and/or negatively impacts human thermal comfort, can be influenced by the structural complexity of the vegetation in both streetscapes and greenspaces (Fig. 2). By structural complexity, we refer to features such as the presence of multiple canopy layers, variation in vegetation volume across vertical strata, and the density and diversity of stems. Importantly, these structural attributes can be achieved through a variety of vegetation compositions, highlighting the importance of managing for structure as a central objective in urban greenspace planning. Vegetation with high structural complexity in a greenspace (greater canopy volume, canopy density or combined tree, shrub and ground layers) is likely to increase shade and transpiration cooling but could also increase relative humidity and reduce windspeed. Although increased structural complexity may raise air humidity and reduce wind speeds, the net effect is a substantial reduction in heat stress levels for humans or animals within these greenspaces (Fig. 2f). Alternatively, vegetation of low structural complexity, such as sparse trees over grass or street trees over hard paving might provide good shade and transpiration cooling but still allow some air movement (Fig. 2b, e). The net effect being an intermediate level of heat stress as compared to high levels of heat stress in open streetscapes without vegetation (Fig. 2a), or greenspace with ground grass cover only (Fig. 2d). Moreover, the net effect of vegetation complexity on microclimate and human thermal comfort is strongly influenced by urban morphological heterogeneity, particularly variations in building configuration and density, which shape both local climate dynamics and thermal exposure (Rahman et al. 2020a).

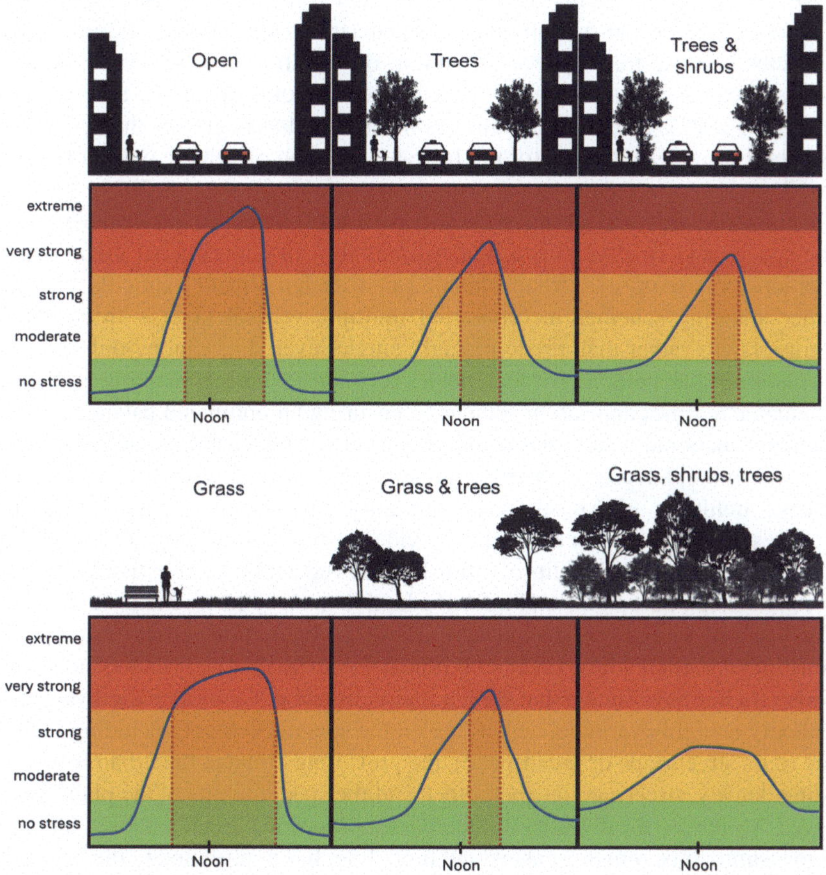

Fig. 2 Conceptual diagram of increasing structural complexity of vegetation in urban streetscapes (upper) and greenspace (lower) and the likely impact upon diurnal heat stress levels (UTCI) on a warm and sunny day. In greenspaces, increasing structural complexity emphasizes the presence of multiple canopy layers and stems of varying diameter and age

Increasing the structural complexity of vegetation in streetscapes is more complex because of the issues of pedestrian access, safety and line-of-sight. However, adding grass and herb ground cover and shrubs, especially in street tree footwells, is gaining greater acceptance (Lička and Furchtlehner 2019). The implications for local microclimate and pedestrian thermal comfort are hard to predict. It is likely that the impact of heat stress levels will be minimal as the shade and transpiration increase will be small, as will the reduction in windspeed. However, if the shrub layer added is more like a continuous hedge with occasional street trees, then the impact on pedestrian thermal comfort may be far greater because shade levels will increase and dispersion (loss) of cooler air under the trees will be reduced. Interestingly, adding a continuous shrub-hedgerow to street trees is promoted as a means to reduce pedestrian exposure to vehicle pollution emissions (Abhijith and Kumar 2021). The reason being that street trees alone can actually increase pedestrian exposure to pollution emitted by vehicles because trees reduce dispersion of pollution up and out of the street canyon and can accumulate higher concentration of pollutants under their canopies (Hang et al. 2023). Adding a continuous hedgerow that is greater than pedestrian head height provides a barrier for the lateral dispersion of vehicle pollution toward sidewalk pedestrians and promotes dispersion upwards (Tran et al. 2022).

To enhance structural complexity in urban environments, particularly within public spaces and streetscapes, a fundamental shift in conventional urban vegetation practices is required. Urban planning and design should explicitly prioritize stratified planting schemes incorporating multiple vertical layers—trees, shrubs, and grass lawns rather than predominantly single-layered or canopy-only systems. Implementing such complexity necessitates revisions to current planting guidelines, maintenance regimes and safety standards, ensuring an optimized balance between ecological functionality, air quality and requirements for visibility and public security. Furthermore, advancing interdisciplinary collaboration among urban planners, landscape architects, ecologists, and community stakeholders is critical. Central to this process is the integration of robust, evidence-based knowledge on the ecological, climatic, and social benefits of multi-layered vegetation, to effectively challenge traditional design paradigms and inform adaptive management practices.

The specific microclimate impacts of increased vegetation structural complexity within cities are still not fully understood and it remains a relatively understudied area. Previous studies have shown that the net microclimate effect of increased vegetation complexity is highly variable, depending on the specific context, including the type of site (e.g., streetscape or greenspace), the prevailing climate, the urban form (e.g., building density, street canyon geometry), and the specific traits of the plant species involved. Moreover, the diversity of methods employed in previous studies, including field measurements, remote sensing and modeling, has contributed to the variability in results observed (Quillet et al. 2010). Therefore, there is a need for systematic and quantitative research to determine the overall impact of multi-layered vegetation structure, such as grass, shrubs, and trees, on microclimate in urban greenspaces and streets when evaluating the effects of vegetation complexity. This will help to identify optimal configurations and thresholds of vegetation complexity to maximize cooling benefits which can support urban streetscapes/greenspaces that improve both urban nature and human thermal comfort in cities.

The aim of this chapter is to explore the microclimate impacts of increased vegetation complexity in both streetscapes and urban greenspaces, and how this knowledge can be incorporated into future assessments and climate-sensitive design. Specifically, we (1) conduct a literature search and draw upon existing empirical case studies to explore the range of potential positive and negative effects of vegetation complexity on urban microclimates, (2) discuss the theoretical implications of background climate, urban form, and plant species characteristics on these complex interactions, and (3) provide recommendations for future research directions that will aid in the development of urban nature-based solutions that support both urban vegetation complexity and human thermal comfort.

2 Methods

2.1 Literature Review

To explore the impact of vegetation complexity on human thermal comfort we first conducted literature searches separately for streetscapes and urban greenspaces using the Scopus database in March 2025. For both searches we applied the same set of keywords related to vegetation and vegetation complexity, combined with terms associated with microclimate and thermal comfort. The searches were limited to English-written and peer-reviewed full-length research articles. For streetscapes, the query used was: TITLE-ABS-KEY ((“urban streetscape” OR “street canyon”) AND (“microclimate” OR “thermal comfort” OR “air temperature” OR “wind speed” OR “relative humidity”) AND (“vegetation” OR “vegetation configuration” OR “multi-layered vegetation” OR “vegetation structure” OR “green” OR “tree” OR “hedge” OR “shrub” OR “grass”)), which yielded 202 peer-reviewed journal articles. For urban greenspaces, the terms “urban green space” OR “urban park” replaced “urban streetscape” in the same search structure, resulting in 516 articles. An additional five relevant articles were manually added to capture important studies not identified by the initial search. All retrieved articles were exported to Zotero for reference management.

To be included, a selected article had to meet the following criteria: (1) it must have assessed one or more forms of vegetation structure (e.g., trees only; grass only; trees + grass; trees + grass + shrub) and (2) it must have reported changes in either Physiological Equivalent Temperature (PET) (Hirashima et al. 2018; Mayer and Höppe 1987) and the Universal Thermal Climate Index (UTCI) (Bröde et al. 2012), both indicators of human thermal comfort. These changes needed to be measured by comparing vegetated settings with unvegetated streetscapes or open greenspaces. Articles were initially screened by reviewing their title and abstract. Those that appeared relevant underwent a full-text assessment to confirm their eligibility. This

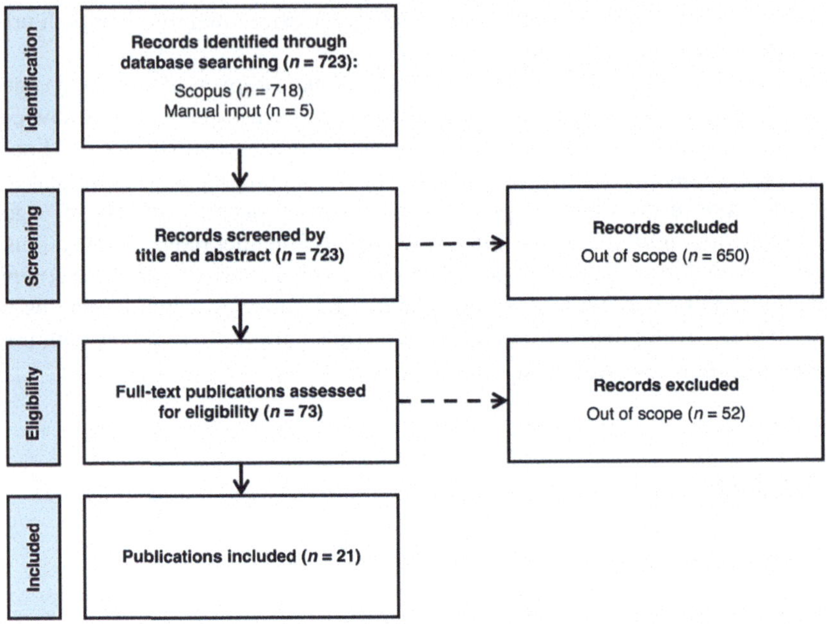

Fig. 3 Flow diagram of the literature search strategy and article selection process for in-depth analysis

process resulted in 21 included articles, with 17 focusing on streetscapes and 4 on greenspaces (Fig. 3).

Key information extracted from the reviewed articles included publication details (i.e., author, year, journal); study location, setting and climate type; vegetation configuration examined, and the methodology used (i.e., field measurements or modeling). Results relating to the changes in human thermal comfort, specifically PET and UTCI, were also recorded. For articles classified under the streetscape category, additional details were collected on canyon geometry (i.e., aspect ratio and canyon orientation) and vegetation characteristics (e.g., tree spacing, crown size and vegetation height).

2.2 *Greenspace and Streetscape Case Study Cities*

To evaluate the influence of vegetation complexity on microclimatic conditions, we reanalyzed data from studies in which detailed in-situ microclimate measurements were collected concurrently, using comparable methods, from either greenspace or streetscape sites of increasing vegetation structural complexity, across contrasting climate zones, including temperate and humid subtropical cities. For streetscape case studies, these included streets or public squares without overhead tree canopy (Open "A"), streets or squares with overhead tree canopy (Trees "B"), and streets or squares with trees, shrubs, and grass (Trees and shrubs "C"). Similarly, greenspace case studies included measurements above simple grass lawns (Grass "A"), grassed areas with overhead tree canopy (Grass and Trees "B"), and sites with grass, shrubs and trees ("C") (Figs. 2 and 4). By selecting case studies that concurrently measured at sites A and B, or even better A, B and C within the same city or suburb, we are better able to understand and test the conceptual framework of how increasing vegetation structure for biodiversity habitat will impact urban microclimate conditions and human thermal comfort.

City case studies were also selected based on the availability of diurnal measurement of key microclimatic variables: air temperature, relative humidity, solar radiation, wind speed, and black globe temperature. These parameters are vital in assessing the energy balance and the thermal comfort in outdoor urban environments. A combination of portable and fixed weather stations were used in the case studies selected. All the measurements were taken at pedestrian level, between 1.3 and 2.0 m above ground surface. All measurements were taken during the summer months. While measurement campaigns were conducted independently and not synchronized across the three cities, each site's dataset includes either full diurnal cycles or at least daytime monitoring (typically from 10:00 to 18:00), with sampling frequencies and durations aligned to the original study designs.

Relevant city case studies were identified from Hong Kong (China), Melbourne (Australia) and Munich (Germany). Hong Kong was able to provide measurement from greenspace sites A, B and C (Figs. 4 and 5). Melbourne was able to provide measurement from greenspace sites A and B, and streetscape sites A and B. Munich was able to provide measurements from greenspace sites A, B and C, and streetscape sites A, B and C (Figs. 4 and 5).

Hong Kong, a high-density metropolis with a population of approximately 7.3 million, represents a humid subtropical climate (Cwa). It experiences a long-term mean annual temperature of 23.3 °C and a significantly higher annual precipitation total of approximately 2398.5 mm (Hong Kong Observatory 2025). The study site for Hong Kong is the Hong Kong Golf Club in Fanling, which is located 15 km north of the city center and occupies an area of approximately 1.7 km^2. The monitoring period was from 23-06-2016 to 25-06-2016. The local climate zone of this greenspace is LCZ B "scattered trees". This greenspace offers a great opportunity to study the impacts of vegetation structure on microclimate because it has a variety of landscapes. Meteorological measurements were conducted at three locations to represent three vegetation structures: HK-Green-A: grass, HK-Green-B: grass and trees, HK-Green-C: grass, shrubs, and trees (Fig. 4). HK-Green-A is located at the edge of a fairway with carpetgrass (*Axonopus compressus*) being the dominant species.

HK-Green-B is a cultivated tree strip adjacent to fairways with carpet grass. The tree heights were approximately 10 m and their canopies did not fully overlap. HK-Green-C is a multi-story natural woodland with various tree species. The tree canopies fully overlapped, allowing only sparse shrub cover and little grass cover. The maximum tree height was approximately 15 m. The sky view factors of HK-Green-A, HK-Green-B, and HK-Green-C were 0.74, 0.25, and 0.08, respectively. All three locations were unirrigated.

Fig. 4 The microclimate of grass only (**a**), grass and trees (**b**) and grass, trees and shrubs (**c**) measured in detail in greenspaces in Hong Kong (China), Melbourne (Australia) and Munich (Germany)

Melbourne has a population of around 4.9 million and a temperate oceanic climate (Cfb). The city records a mean annual temperature of 14.8 °C and annual rainfall of approximately 648 mm (Bureau of Meteorology 2025). The greenspace study site for Melbourne is a research facility (0.07 km^2) of the university of Melbourne. It is located approximately 10 km east of Melbourne's Central Business District.

Fig. 5 The microclimate of open streetscapes (**a**), overhead trees in streets (**b**) and overhead trees and under-canopy shrubs in streets (**c**) were measured in detail across streetscapes in Melbourne (Australia) and Munich (Germany)

The study period was from 16-03-2024 to 18-03-2024. The local climate zone of this greenspace is LCZ B "scattered trees". Two identical 6 × 6 m (36 m^2) plots were used to install two vegetation structures—MEL-Green-A: grass, MEL-Green-B: grass and trees (Fig. 5). MEL-Green-A was irrigated turf (*Pennisetum clandestinum*). The plots were built to mimic the environment of private backyards. MEL-Green-B consisted of four trees (*Syzygium floribundum*) that were approximately 3 m tall and planted at the vertices of a 2 m^2. A 1.8 m tall shade cloth was used to enclose the plots to reduce mixing of air between the plots and the surroundings. Both plots were irrigated 4 mm d^{-1}.

Melbourne also provides a streetscape case study in a residential street in the Richmond suburb planted with *Platanus* × *acerifolia* trees. Clifton Street is north–south-oriented with a canopy cover of approximately 42% (Fig. 5). Microclimate data were collected on four separate days: 18 December 2013, 7 January 2014, 10 January 2014, and 13 January 2014.

Munich, the third-largest city in Germany with a population of 1.6 million, represents a temperate oceanic climate (Cfb). It has a long-term mean annual temperature of 10.1 °C and receives about 939.7 mm of precipitation annually (DWD 2021). The greenspace sites in Munich included two locations: Nordbad (MUC-Green-A, MUC-Green-B) and in Akademie (MUC-Green-C) (Fig. 4). Nordbad is the park area surrounding an indoor swimming pool, characterized by grass and tree cover dominated by *Tilia cordata* trees. These trees had an average diameter at breast height (dbh) of 45 cm and a mean height of 17.3 m. In contrast, Akademie is a more complex park with a multi-layered vegetation structure, predominantly composed of *Acer platanoides*. Trees at this site had an average dbh of 30.6 cm and height of 18 m. Measurements were carried out over at least three days at each site between 9 July and 23 August 2023. The streetscape study sites in Munich included Alter Hof (MUC-Street-A), Elisabethstraße (MUC-Street-B), and Zenettiplatz (MUC-Street-C). Alter Hof is a highly sealed public square surrounded by buildings. Elisabethstraße is a residential, east–west-oriented street planted with *Acer platanoides*, with trees averaging 30 cm in dbh and 13 m in height. Zenettiplatz, a public square with a multi-layered vegetation configuration, is planted with *Robinia pseudoacacia* species and encircled by approximately 1 m high shrubs. The average dbh and height of the Robinia trees were 25 cm and 14.5 m, respectively. Measurements at Alter Hof and Elisabethstraße were conducted over three days between 7 and 23 July 2023, while data collection at Zenettiplatz took place on 13, 18, and 31 July 2024.

2.3 Data Analysis for the Case Studies

Using the collected data, we calculated the mean radiant temperature (T_{mrt}), which describes the radiant heat exchange between the person and the surroundings and is an influential parameter of outdoor thermal comfort. T_{mrt} was calculated using Eq. (1) (ISO 1998):

$$T_{\mathrm{mrt}} = \sqrt[4]{\left(T_g + 273.15\right)^4 + \frac{1.1 * 10^8 * V_a^{0.6}}{\varepsilon * D^{0.4}} * \left(T_g - T_a\right)} - 273.15 \qquad (1)$$

where T_g is the globe temperature (°C), V_a is the wind speed (m/s), T_a is the air temperature (°C), D is the diameter of the globe (= 150 mm), and ϵ is the globe emissivity (= 0.95)

As a quantitative measure of human thermal comfort, we use the Universal thermal climate index (UTCI) (Bröde et al. 2012). UTCI is a thermal comfort index that integrates the impacts of air temperature, humidity, mean radiant temperature, and wind speed on human thermal stress. The UTCI value can be categorized in terms of thermal stress according to the UTCI standard assessment scale, ranging from extreme cold stress (UTCI < −40 °C) to extreme heat stress (UTCI > 46 °C). UTCI was calculated using the "rBiometeo" package in R Studio 4.4.2 (R Core Team 2024). For the literature review, along with the UTCI, we also analyzed the Physiological Equivalent Temperature (PET). PET is a thermal comfort index that converts the effects of air temperature, humidity, wind speed, and radiation into an equivalent indoor air temperature, with thermal stress categories ranging from extreme cold stress (PET < 4 °C) to extreme heat stress (PET ≥ 41 °C) (Mayer and Höppe 1987).

For case study cities, we analyzed the influence of vegetation complexity on UTCI. To ensure comparability across sites and cities, we first filtered the datasets to exclude days with overcast or rainy conditions. Only warm, clear, and sunny summer days were retained for analysis. For each selected day, all variables were aggregated to hourly means to smooth short-term fluctuations.

3 Synthesis of Evidence

3.1 Evidence from Recent Relevant Literature

The review of relevant urban vegetation, microclimate and human thermal comfort studies over the last 10 years (since 2015) revealed a greater number of studies on vegetation in streetscapes, than in urban greenspaces (Tables 1 and 2). Furthermore, a greater proportion (two thirds) of studies rely upon computational fluid dynamic modeling (e.g., ENVI-met, PALM) rather than field measurements (one third). Many of these computational fluid dynamic model studies do make use of some measurements for either calibration or validation, but their understanding of vegetation complexity impacts upon street or greenspace microclimate and thermal comfort is determined from comparison of simulated vegetation scenarios.

Of the reviewed papers, only computational fluid dynamic modeling studies truly investigated multi-layer vegetation systems of increasing structural complexity, principally comparing open streetscape or open grass greenspace scenarios to the addition of trees (trees and grass), shrubs, or trees and shrubs (Dong et al. 2023; Lee et al. 2023; Lobaccaro and Acero 2015; Scheuer et al. 2024; Wu et al. 2025). The field measurement studies were restricted to comparisons of open streetscape or open grass greenspace scenarios with the addition of only trees. These field measurement studies looked at different scenarios of tree structural complexity according to cover percentage, canopy closure, crown size, planting design or species type.

The key findings that emerged from a review of these studies were that the addition of tree structural complexity, for instance, adding trees on the bare streets to streetscapes can reduce the daytime average of human thermal stress by approximately 5.0–6.0 °C UTCI and 2–8.7 °C PET which is sufficient to lower heat stress

Table 1 Urban streetscape studies (2015–2025) using models (WHITE) or field measurements (GREY) to estimate human thermal comfort through PET/UTCI differences between open unvegetated streetscapes as compared to streetscapes with trees, grass or trees and grass.

City/country	Climate (Köppen classification)	Model or measure	Vegetation vs none	ΔPET or UTCI* (°C) Avg	ΔPET or UTCI* (°C) Max	Source
Isfahan, Iran	Semi-arid (BSk)	ENVI-met	- Trees	−5.0		Narimani et al. (2022)
Taiyuan, China	Temperate monsoon (Dwa)	ENVI-met	- Trees	−5.7		Chen et al. (2024)
Roorkee, India	Humid subtropical (Cwa)	ENVI-met	- Trees	−2.4	−6.1	Mohammad et al. (2021)
Hangzhou, China	Subtropical monsoon (Cfa)	ENVI-met	- Trees		−24.0	Zhang et al. (2025)
Prague, Czech Rep	Humid continental (Cfb)	PALM (LES)	- Trees	−5.0*		Janků et al. (2024)
Bangkok, Thailand	Humid tropical (Aw)	ENVI-met	- Trees		−8.6	Srivanit and Jareemit (2020)
Freiburg, Germany	Temperate oceanic (Cfb)	ENVI-met	- Trees	−4.3		Lee et al. (2020)
Xi'an, China	Humid subtropical (Cwa)	ENVI-met	- Trees	−4.8	−10.4	Yang et al. (2018)
Huế, Vietnam	Tropical monsoon (Am)	ENVI-met	- Trees - Grass - Trees + Grass		−1.88* −0.09* −1.96*	Scheuer et al. (2024)
Jeju, South Korea	Humid subtropical (Cfa)	EMVI-met	- Trees - Trees + Grass	−6.2 −6.3	−16.0 −16.1	Lee et al. (2023)
Bilbao, Spain	Temperate oceanic (Cfb)	ENVI-met	- Grass - Trees + Grass	−4.2 −6.7		Lobaccaro and Acero (2015)
European cities	–	ENVI-met	- Trees - Grass	−8.7 −1.1		Wu et al. (2025)
Melbourne, Australia	Warm Med. (Cfb)	Measured	- Trees		−4.4 to −4.6	Sanusi et al. (2016)
Melbourne, Australia	Warm Med. (Cfb)	Measured	- Trees	−4.8 to −5.3		Sanusi et al. (2017)
Würzburg, Germany	Temperate oceanic (Cfb)	Measured	- Trees	−1.6 to −3.0	−4.0 to −11.0	Rahman et al. (2020a)

(continued)

Table 1 (continued)

City/country	Climate (Köppen classification)	Model or measure	Vegetation vs none	ΔPET or UTCI* (°C) Avg	ΔPET or UTCI* (°C) Max	Source
Melbourne, Australia	Warm Med. (Cfb)	Measured	- Trees	−6.0*		Coutts et al. (2016)
Shenyang, China	Mid-temp. continental (Dwa)	Measured	- Trees	−2.1		Miao et al. (2023)

Table 2 Urban greenspace studies (2015–2025) using models (WHITE) or field measurement (GREY) of microclimate to estimate PET/UTCI differences between open greenspace with grass as compared to greenspace with trees, shrubs, or trees and shrubs

City/country	Climate (Köppen classification)	Model or measure	Vegetation vs grass	ΔPET or UTCI* (°C) Avg	ΔPET or UTCI* (°C) Max	Source
Perth, Australia	Mediterranean (Csa)	ENVI-met	- Trees		−17.0*	Barghchi et al. (2024)
Hefei, China	Humid subtropical (Cfa)	ENVI-met	- Shrub - Trees - Trees + Shrub		−0.72 −7.32 −7.44	Dong et al. (2023)
Mianyang, China	Humid subtropical (Cwa)	Measured	- Trees (80% canopy)	−23.2		Liu et al. (2023)
Guangzhou, China	Humid subtropical (Cfa)	Measured	- Trees	−3.5		Wang et al. (2018)

experienced by pedestrians from for examples "strong heat stress" to "moderate heat stress". The addition of a grassy ground level to these street trees only led to a small (0.1–2.0 °C PET or UTCI) further reduction in human heat stress. There was far greater variation in the estimated reductions in maximum human thermal stress at particular times of the day from an increase in street vegetation structural complexity. Some studies modeled up to a 24.0 °C thermal stress reduction (Zhang et al. 2025) or measured up to a 11.0 °C thermal stress reduction (Rahman et al. 2020a) typically during early to mid-afternoon hours when thermal stress and ambient temperatures peak, whereas other studies measured maximum heat stress reductions of only 2.0–6.0 °C (Mohammad et al. 2021; Sanusi et al. 2016; Scheuer et al. 2024).

Another key finding was that an increase in vegetation structural complexity in greenspaces, beyond turfgrass alone, was just as impactful upon human thermal comfort as adding vegetation to streetscapes. Open greenspace can experience levels of heat stress similar to those in open streetscapes of grayspaces on hot summer days (Barghchi et al. 2024) and adding trees and shrubs will lead to reductions in human heat stress of between 3.0 and 23.0 °C PET or UTCI (Dong et al. 2023; Liu et al. 2023; Wang et al., 2023).

3.2 *Case Studies*

3.2.1 Greenspaces

Hong Kong

In Hong Kong, between 14:00 and 16:00, HK-Green-B and HK-Green-C recorded lower air temperatures and mean radiant temperatures than HK-Green-A, leading to reduced UTCI values (Table 3). Around 12:00, HK-Green-A and HK-Green-B reached "very strong heat stress" levels (UTCI = 38.0–45.9 °C), while HK-Green-C remained within the "strong heat stress" category (UTCI = 32.0–37.9 °C). The results indicate that greater vegetation complexity, particularly overlapping tree and shrub layers, effectively mitigated midday thermal stress in this subtropical climate (Fig. 6). The comparison of the microclimate across the three case study locations indicated that when tree and shrub canopies overlap the greater interception of incoming solar radiation reduces air temperature, mean radiant temperature and human heat stress.

Melbourne

In Melbourne, between 14:00 and 16:00, MEL-Green-A and MEL-Green-B exhibited similar air temperatures, relative humidities, and wind speeds, but MEL-Green-B recorded substantially lower mean radiant temperature and UTCI (Table 3). During the daytime, MEL-Green-B consistently maintained lower UTCI values than MEL-Green-A (18.4 °C and 5.0 °C, respectively), although at night MEL-Green-A exhibited slightly lower UTCI. These results highlight the role of increased vegetation complexity in reducing daytime thermal stress (Fig. 6). The comparison between MEL-Green-A and MEL-Green-B suggested that, in the afternoon, small trees are

Table 3 Microclimate and human thermal comfort (UTCI) differences between grass with trees and grass only (B–A) and between grass, trees, and shrubs and grass only (C–A) in Hong Kong, Melbourne and Munich case studies

City		T_a (°C)	RH (%)	W_s (m/s)	T_{mrt} (°C)	UTCI (°C)
Hong Kong	B–A	−1.0	+1.1	+0.1	−3.5	−1.8
Hong Kong	C–A	−2.0	+7.6	−0.1	−5.2	−3.5
Melbourne	B–A	+2.8	+0.6	0.0	−20.7	−6.5
Melbourne	C–A	–	–	–	–	–
Munich	B–A	−3.7	+9.0	−0.3	−14.5	−5.9
Munich	C–A	−5.2	+18.5	−0.3	−18.0	−7.8

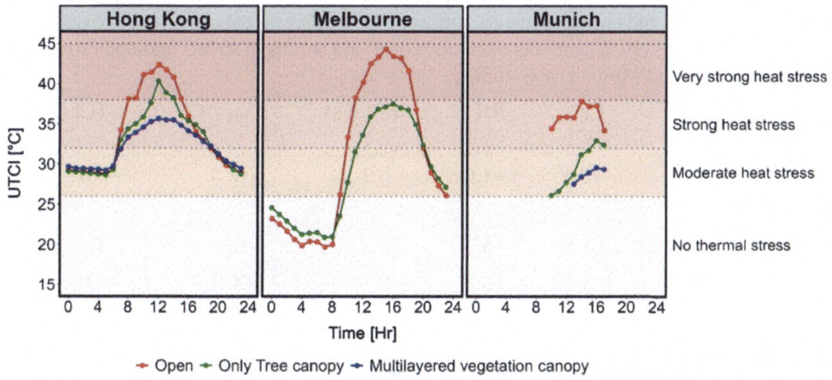

Fig. 6 Hourly mean UTCI values for greenspace sites in Hong Kong, Melbourne and Munich, grouped by vegetation configuration

not effective in reducing air temperature but effective in reducing mean radiant temperature and heat stress.

Munich

In Munich, between 14:00 and 16:00, MUC-Green-A exhibited the highest UTCI values (36.0 °C, "strong heat stress"), while MUC-Green-B and MUC-Green-C, with more complex vegetation structures, recorded substantially lower UTCI values (29.0 °C and 28.8 °C, respectively), corresponding to "moderate heat stress" (Table 3). The comparison across sites demonstrates that increasing vegetation complexity consistently reduced thermal stress, with multi-layered canopies particularly effective at suppressing midday UTCI peaks (Fig. 6). All these case studies suggested that heat stress can be reduced by tree shade even though trees may reduce wind speed substantially.

3.2.2 Streetscape

Melbourne

In Melbourne, between 14:00 and 16:00, the tree-only configuration (MEL-Street-B) reduced air temperature and mean radiant temperature (Table 4), which significantly reduced heat stress from strong to moderate in the afternoon (Fig. 7). The open configuration (MEL-Street-A) experienced a steep and sustained increase in UTCI from morning through mid-afternoon, whereas the tree-only site maintained consistently lower values throughout the day and remained below the strong heat stress threshold.

Table 4 Microclimate and human thermal comfort (UTCI) differences between tree cover in streets and in open streetscapes (B–A), and between tree cover and shrubs in streets and open streetscapes (C–A) in Melbourne and Munich case studies

City		T_a (°C)	RH (%)	W_s (m/s)	T_{mrt} (°C)	UTCI (°C)
Melbourne	B–A	−0.6	+1.6	−0.3	−18.4	−5.0
Melbourne	C–A	–	–	–	–	–
Munich	B–A	+0.4	+5.3	0.0	−16.6	−4.3
Munich	C–A	−1.6	+6.8	−0.1	−20.1	−6.0

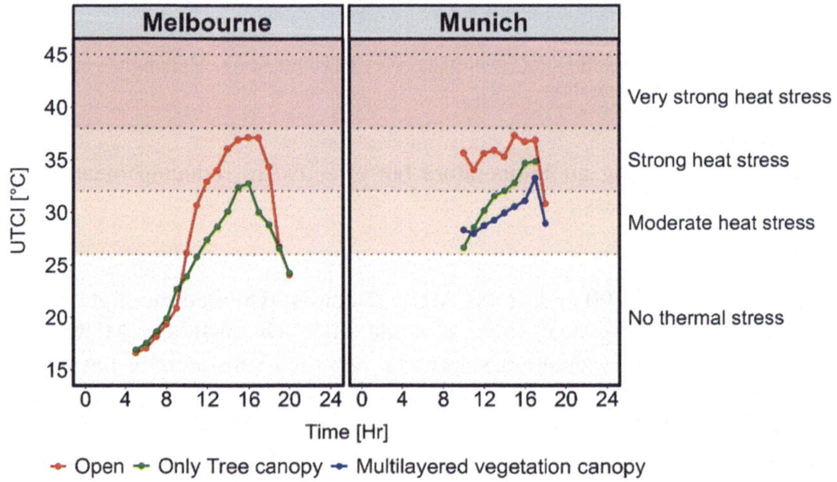

Fig. 7 Hourly mean UTCI at streetscape sites in Melbourne and Munich by vegetation configuration (multilayered canopy in Munich includes trees, shrubs, and lawns)

Munich

In Melbourne, between 14:00 and 16:00, the impacts of the tree-only configuration (MUC-Street-B) and multi-layered vegetation configuration (MUC-Street-C) on air temperature was mixed. MUC-Street B increased air temperature, whereas MUC-Street-C reduced it (Table 4). However, both MUC-Street-B and MUC-Street-C reduced mean radiant temperature and heat stress effectively. The open configuration (MUC-Street-A) sustained a strong heat stress from 08:00 to 16:00, whereas MUC-Street-C stayed within moderate heat stress for most of the time in this period (Fig. 7).

3.3 Microclimatic Assessment Results at Peak Thermal Stress Hours (14–16 h)

Across both streetscapes and greenspaces, increased vegetation complexity consistently reduced thermal loads during peak afternoon hours between 14 and 16 h (Tables 3 and 4). The greatest UTCI reductions were observed in greenspaces compared to streetscapes. In Melbourne greenspaces, only tree canopies (B–A) lowered UTCI by 6.5 °C, and T_{mrt} by 20.7 °C. In case of multi-layered vegetation (C–A), in Munich greenspaces, the UTCI reduced by 7.8 °C and T_{mrt} by 18.0 °C between open and multilayered sites. Hong Kong greenspaces followed a similar trend, though the magnitude of reduction was smaller: UTCI was reduced by 3.5 °C, and T_{mrt} by 5.2 °C from open to multilayered conditions.

In streetscapes, Melbourne showed large absolute difference in T_{mrt} (B–A: −18.4 °C) and a UTCI reduction of 5.0 °C. For Munich streetscapes, the UTCI difference reached 6.0 °C, while T_{mrt} fell by 20.1 °C between configurations C and A. In most contexts, air temperature reductions were modest (<2 °C) compared to the more substantial differences observed in radiant temperature and UTCI. Relative humidity consistently increased with vegetation complexity, particularly in Munich greenspaces, where RH rose by 18.5% (C–A). Wind speed differences remained minimal across all settings.

4 Integration of Findings and Broader Implications

The literature review and case study analysis demonstrate that urban nature-based solutions that increase the structural complexity of vegetation for biodiversity habitat will change local microclimate conditions and thereby reduce levels of thermal stress of humans and animals. Increases in both tree and shrub structural complexity are associated with higher plant diversity and quality habitat for wildlife (Threlfall et al. 2017, 2016), suggesting that promoting tree and shrub structural complexity could improve both human thermal comfort and support diverse habitats for nature within cities.

The findings from our case studies complement and extend those identified in the literature review (Tables 1, 2, 3 and 4; Figs. 6 and 7). Our empirical measurements showed that tree-only configurations in streetscapes reduce PET/UTCI by an average of 4–5 °C, while the addition of grass or shrubs provides further meaningful improvements, contributing up to an additional 2 °C reduction in thermal stress. These additive effects are more pronounced in urban greenspaces compared to streetscapes. The greater additive cooling effect of shrubs in greenspaces compared to streetscapes can be attributed to the relative contribution of vegetation layers: in greenspaces, the introduction of a shrub layer constitutes a substantial increase in total vegetative volume and surface area for evapotranspiration and shading, whereas in streetscapes, the limited addition of small hedges or understory shrubs beneath existing street trees represents a comparatively minor modification to the overall vegetation structure and microclimatic function.

In contrast to the dominant reliance on computational fluid dynamics (CFD) modeling in the literature, our field-based approach provides real-world validation and reveals that actual cooling effects may be lower than model projections in complex urban contexts due to interactions with ambient conditions (e.g., ready advection from the surrounding environment). For instance, our Melbourne streetscape data showed around ~5 °C UTCI reduction for adding trees on streets, compared to modeled maximums exceeding 6–10 °C in other cities, suggesting the importance of validating models with local, empirical data. The cooling impacts of urban vegetation are significantly shaped by background climate, as the relative importance of evapotranspiration and shading varies across climatic zones; humid climates enhance transpiration cooling, while in arid climates, shading becomes the dominant mechanism (Li et al. 2024; Rahman et al. 2024).

4.1 Contributions to the Understanding of Microclimate Impacts of Multi-layered Vegetation

Our case studies along with synthesis of literature encompassing horizontal and vertical layering of grass, shrubs, and trees presents a complex mix of benefits and drawbacks to advance our understanding in modulating urban heat. Incorporating structurally complex vegetation into streetscapes and greenspaces has been shown to significantly reduce daytime human thermal stress. For example, our case studies reported reductions in T_{mrt} by up to 20.7 °C and in the UTCI by up to 7.8 °C when compared to open streetscape or open grass greenspaces. These reductions are sufficient to lower pedestrian heat stress levels from "strong heat stress" to "moderate" or even "slight heat stress" during midday and early afternoon hours, when solar radiation and ambient temperatures peak; however, it should be acknowledged that at night, the influence of dense vegetation can reverse, potentially leading to localized heat retention due to reduced longwave radiation loss from the surrounding built environment and restricted air circulation.

A key insight emerging from our integrated analysis is the context-specific and multifaceted influence of vegetation structural complexity on net microclimatic outcomes. While previous literature emphasizes a generally positive net effect, our data emphasize that trade-offs exist, particularly regarding humidity and airflow. For example, in high-humidity environments like Hong Kong, increased vegetation complexity compounded local moisture levels, which may reduce thermal comfort during humid heatwaves. While increasing air humidity might be beneficial in dry climates, increased relative humidity can hinder sweat evaporation and human comfort in already humid environments. In our case, we found over 18% increase in relative humidity in complex greenspaces in Munich. Dense vegetation can reduce wind speed, limit convective cooling and pollutant dispersion. This effect was observed in Munich and Melbourne's streetscapes as well as the greenspaces of Munich where vegetation decreased wind speed and increased air stagnation. Trees in street canyons may trap vehicular emissions, but adding shrubs can reroute pollutants upwards, as demonstrated in studies from high-traffic streets (Islam et al. 2024; Vos et al. 2013). Jeanjean et al. (2016) reported small reductions in particulate matter ($PM_{2.5}$) concentrations of 0.6% for grass and 2.8% for trees. Dense tree canopies can trap pollutants under the canopy by restricting dispersion and mixing with cleaner air from above, contributing to pollutant build-up near ground level. Hang et al. (2023), through modeling, found that trees in a street canyon lowered wind speed by up 70%, leading to reduced nitrogen dioxide dispersion and a 66% increase in its concentration on the leeward side. Addressing a wider range of vegetation structures, a study in Shenzhen, China, combining computational and physical modeling, reported that all examined vegetation configurations–trees, shrubs and tree-shrub combinations–increased PM concentrations at 1.5 m compared to no vegetation within a street canyon, and among these, shrubs alone had the least negative impact on particle dispersion, with PM concentration 15–20% lower than those of the other two configurations (Wu et al. 2021).

4.2 *Theoretical Implications: Climate, Urban Form, and Plant Species*

Using ENVI-met simulations in a streetscape in Jeju, South Korea, Lee et al. (2023) demonstrated that adding grass to areas with existing tree cover resulted in only a marginal PET reduction of 0.1 °C - from 6.2 to 6.3 °C. In contrast, Lobaccaro and Acero (2015) reported a more substantial PET reduction of 2.5 °C, from 4.2 to 6.7 °C, following similar interventions in Bilbao, Spain. Findings from Bilbao align closely with our case study in Munich, Germany, where the addition of an extra vegetation layer increased PET reduction by 1.7 °C from 4.3 to 6 °C. In Hong Kong's greenspaces, the overall capacity of vegetation to reduce heat stress was lower but improvement by increasing vegetation structural complexity was more similar to Munich, with PET reduction by 1.7 °C increasing from 1.8 to 3.5 °C. These varying outcomes highlight two key points: first, that current microclimate models are constrained by the limitations of our existing knowledge on vegetation–atmosphere interactions; and second, that the lack of standardized protocols, particularly the absence of rigorously designed paired datasets with consistent control scenarios poses challenges for accurately quantifying the thermal benefits of vegetation across different urban contexts for comparative analysis. Our study reinforces that vegetation complexity must be viewed not as a simple additive variable, but as a multidimensional construct involving plant traits (e.g., LAI, canopy closure), spatial arrangement (e.g., vertical layering, density), and contextual modifiers (e.g., canyon geometry, background climate). These findings echo and build upon the theoretical frameworks discussed by Wang et al. (2023), which underscore the interplay between vertical structure and local microclimate processes.

Cooling benefits vary with climate. In tropical climates (e.g., Bangkok), shade is more important than transpiration due to already high humidity (Srivanit and Jareemit 2020). In temperate cities like Melbourne, transpiration also plays an important role. Case studies showed stronger UTCI reductions in temperate and subtropical cities compared to humid tropical cities, highlighting the climate dependency of vegetation cooling benefits. At the same time urban form plays a big role in modulating cooling benefits; streets with low height-to-width ratios allow better sunlight penetration and airflow. In Melbourne, east–west aligned streets with trees achieved greater cooling (up to 5.3 °C reduction in UTCI). Similarly, canyon geometry plays a role, for instance, in Munich, narrower streets reduced wind movement, amplifying the humidifying and pollutant-retention effects of dense vegetation. Moreover, plant traits such as Leaf Area Index (LAI), canopy density, and transpiration rates influence thermal regulation. For instance, *Tilia cordata* in Würzburg, Germany reduced PET up to 11 °C compared to only up to 4 °C under the shade of *Robinia pseudoacacia* (Rahman et al. 2020a).

Plant selection will be integral for improving human thermal comfort and nature in cities through enhancing vegetation complexity. Tree diversity in urban greenspaces have been shown to increase bird diversity and habitat for microbats and arboreal mammals (Threlfall et al. 2016). Shrub cover can provide important habitat for birds, invertebrates and ground dwelling invertebrates (Threlfall et al. 2017). However, not all tree species are equal, with some species providing more diverse habitat resources than others. For example, increasing tree and shrub vegetation complexity through selecting indigenous plant species will be crucial for supporting local biodiversity and human thermal comfort simultaneously (Mata et al. 2021; Tan et al. 2022). Plant selection will be an important element for increasing vegetation structural complexity so that both human thermal comfort and biodiversity habitat enhancement. It will also be important that urban nature-based solutions that increase vegetation complexity are designed to enhance connectivity between greenspaces.

4.3 Implications for Urban Planning and Nature-Based Solutions

From a practical perspective, our findings have significant implications for urban planners and landscape architects. Incorporating multi-layered vegetation, particularly, combinations of trees and shrubs is an effective nature-based solution for reducing urban heat stress. However, implementation must be tailored to specific urban contexts. For example: in streetscapes, adding low-level shrubs in combination with trees (especially in footwells or median strips) can enhance cooling and reduce pollutant exposure, but care must be taken to avoid overly restricting airflow and visibility. Whereas, in greenspaces, maximizing vertical complexity through a diverse canopy and shrub layer delivers both thermal comfort and biodiversity benefits, particularly when native species are prioritized.

Importantly, our findings highlight that grass alone provides limited contributions to thermal comfort, with some studies indicating that while grass evapotranspiration and reduced ground heat flux can alleviate urban heat (Barghchi et al. 2024; Lee and Jim 2019; Rahman et al. 2019), their effect on thermal comfort is limited. While the studies identified in our literature review only focused on turf grass, grassland species and herbaceous cover can be important habitat for wildlife (Pithon et al. 2021). This suggests that urban nature-based solutions that focus on improving the vegetation complexity of the grass or herbaceous vegetation layer may improve biodiversity habitat but not human thermal comfort. Therefore areas that aim to enhance grassland diversity could be implemented in areas of urban greenspaces that are further away from where people congregate, and where providing human thermal comfort is less important, such as further away from paths and picnic areas.

Overall, this study supports a shift toward integrated design thinking, where vegetation complexity is calibrated to balance human thermal comfort, biodiversity, and urban functionality. Future city cooling strategies must consider not only vegetation presence, but its form, function, and interaction with the surrounding built environment. However, for a broader and more precise understanding, field-based, long-term studies on fully complex vegetation structures (trees + shrubs + herbs) are needed to capture diurnal and seasonal variability in cooling benefits. While our analysis concentrated on maximum heat stress situations, nighttime and seasonal dynamics were not addressed and warrant further investigation. Evaluating vegetation combinations and configurations across diverse urban contexts using standardized protocols and thermal comfort indices (e.g., UTCI, PET) is equally important. Given the challenges of generating fully experimental results across different urban settings, computational approaches such as CFD modeling are essential; however, developing hybrid methodologies that integrate ENVI-met simulations with empirical field measurements is critical for producing robust, scalable findings applicable across climates and urban morphologies. Additionally, testing vegetation complexity using native species is vital for enhancing ecological value without compromising human thermal comfort, while simultaneously addressing social dimensions of access, safety, and maintenance in the design of multilayered streetscape vegetation.

5 Conclusion

This chapter underscores the complex, context-dependent role that multi-layered vegetation plays in modulating urban microclimates. While increased vegetation complexity through combinations of trees, shrubs, and grasses generally reduces human thermal stress, the magnitude and nature of these benefits vary according to vegetation structure, background climate, urban form, and plant species traits.

Empirical findings from our case studies confirm and extend the literature review evidence, highlighting that real-world impacts often differ from modeled predictions due to local variables such as street geometry, species characteristics, and microclimatic interactions. Notably, the thermal benefits of complex vegetation are more consistently pronounced in greenspaces than in streetscapes, where airflow and pollutant dispersion complicate outcomes.

These findings carry significant implications for urban planning. Nature-based solutions that optimize vegetation complexity not only support urban cooling but also enhance biodiversity and ecosystem connectivity, especially when designed with indigenous species and human use patterns in mind. Future research should explore thresholds and permutations of vegetation layering that maximize cooling without undue trade-offs in humidity or air quality, and test design strategies across diverse urban contexts and climate zones.

Acknowledgements Thanks go to the German Research Foundation (Deutsche Forschungsgemeinschaft) for providing funds for the projects 437788427—RTG 2679. The Hong Kong microclimatic study was supported by the Hong Kong Golf Club and the Research Matching Grant of the Hong Kong Research Grants Council.

References

Abdel-Ghany A, Al-Helal I, Shady M (2013) Human thermal comfort and heat stress in an outdoor urban arid environment: a case study. Adv Meteorol 2013(1):693541

Abhijith K, Kumar P (2021) Evaluation of respiratory deposition doses in the presence of green infrastructure. Air Qual Atmos Health 14(6):911–924

Armson D, Stringer P, Ennos A (2012) The effect of tree shade and grass on surface and globe temperatures in an urban area. Urban For Urban Green 11:245–255

Barghchi M, Grace B, Edwards N, Bolleter J, Hooper P (2024) Park thermal comfort and cooling mechanisms in present and future climate scenarios. Urban For Urban Green 101:128533

Bröde P et al (2012) Deriving the operational procedure for the Universal Thermal Climate Index (UTCI). Int J Biometeorol 56:481–494

Bureau of Meteorology (2025) Climate Data Online, Bureau of Meteorology, Australia

Chen X et al (2024) Integrated effect of aspect ratio and tree spacing on pedestrian thermal comfort of street canyon. Int J Biometeorol 68(10):2115–2131

Coutts AM, White EC, Tapper NJ, Beringer J, Livesley SJ (2016) Temperature and human thermal comfort effects of street trees across three contrasting street canyon environments. Theoret Appl Climatol 124:55–68

Dodman D et al (2022) Cities, settlements and key infrastructure. In: Climate change 2022: impacts, adaptation, and vulnerability. contribution of working group II to the sixth assessment report of the intergovernmental panel on climate change

Dong D, Yu T, Yi L, Wusi Z, Kangkang G (2023) Numerical simulation of effect of vegetation configuration on human thermal comfort. J Environ Occup Med 40(8):900–909

DWD (2021) Deutscher Wetterdienst

Esperón-Rodríguez M et al (2025) Urban heat in global cities and the role of nature-based solutions in mitigating future climate risks. Environ Res: Clim

Fanger PO (1970) Thermal comfort. Analysis and applications in environmental engineering

Garrard GE, Williams NSG, Mata L, Thomas J, Bekessy SA (2018) Biodiversity sensitive urban design. Conserv Lett 11(2):e12411

Hang J et al (2023) Numerical investigation of the impact of urban trees on O3–NOx–VOCs chemistry and pollutant dispersion in a typical street canyon. Atmos Environ 311:119998

Harris V, Dave K, Hahs AK, Threlfall CG (2018) Green space context and vegetation complexity shape people's preferences for urban public parks and residential gardens. Landsc Res 43(1):150–162

Hirashima SQDS, Katzschner A, Ferreira DG, Assis ESD, Katzschner L (2018) Thermal comfort comparison and evaluation in different climates. Urban Clim 23:219–230

Hong Kong Observatory (2025) Monthly meteorological normals for Hong Kong. Hong Kong Observatory, Hong Kong

Huang K, Li X, Liu X, Seto KC (2019) Projecting global urban land expansion and heat island intensification through 2050. Environ Res Lett 14(11):114037

Islam A et al (2024) Impact of urban green spaces on air quality: A study of PM10 reduction across diverse climates. Sci Total Environ 955:176770

ISO (1998) Ergonomics of the thermal environment – Instruments for measuring physical quantities

Iungman T et al (2023) Cooling cities through urban green infrastructure: a health impact assessment of European cities. The Lancet 401(10376):577–589

Janků Z et al (2024) Towards climate-responsible tree positioning: detailed effects of trees on heat exposure in complex urban environments. Urban For Urban Green 101:128500

Jeanjean AP, Monks PS, Leigh RJ (2016) Modelling the effectiveness of urban trees and grass on PM2. 5 reduction via dispersion and deposition at a city scale. Atmos Environ 147:1–10

Lai D, Liu W, Gan T, Liu K, Chen Q (2019) A review of mitigating strategies to improve the thermal environment and thermal comfort in urban outdoor spaces. Sci Total Environ 661:337–353

Lee H, Lim H, Park S (2023) Quantitative assessment of green coverage changes under the human-biometeorological perspective: A simulation case study in Jeju, Republic of Korea. Sustain Cities Soc 97:104734

Lee LSH, Jim CY (2019) Urban woodland on intensive green roof improved outdoor thermal comfort in subtropical summer. Int J Biometeorol 63(7):895–909

Lee H, Mayer H, Kuttler W (2020) Impact of the spacing between tree crowns on the mitigation of daytime heat stress for pedestrians inside EW urban street canyons under Central European conditions. Urban For Urban Green 48:126558

Lička L, Furchtlehner J (2019) Changing streets: individual actions, large-scale measures, and ambitious urban climate goals. In: Rinaldi T (ed) Urban landscapes in high-density cities, pp 100–116

Li H et al (2024) Cooling efficacy of trees across cities is determined by background climate, urban morphology, and tree trait. Commun Earth Environ 5(1):754

Liu Y et al (2023) A comparative study of the influences of park physical factors on summer outdoor thermal environment, a pilot study of Mianyang, China. Nat-Based Solutions 4:100083

Lobaccaro G, Acero JA (2015) Comparative analysis of green actions to improve outdoor thermal comfort inside typical urban street canyons. Urban Clim 14:251–267

Ma S et al (2018) Evaluating the effectiveness of mitigation options on heat stress for Sydney, Australia. J Appl Meteorol Climatol 57(2):209–220

Mata L et al (2021) Indigenous plants promote insect biodiversity in urban greenspaces. Ecol Appl 31(4):e02309

Mayer H, Höppe P (1987) Thermal comfort of man in different urban environments. Theoret Appl Climatol 38:43–49

McDonald RI, Kareiva P, Forman RT (2008) The implications of current and future urbanization for global protected areas and biodiversity conservation. Biol Cons 141(6):1695–1703

Miao C et al (2023) Coupling outdoor air quality with thermal comfort in the presence of street trees: a pilot investigation in Shenyang, Northeast China. J For Res 34(3):831–839

Mohammad P et al (2021) Evaluating the role of the albedo of material and vegetation scenarios along the urban street canyon for improving pedestrian thermal comfort outdoors. Urban Clim 40:100993
Narimani N, Karimi A, Brown RD (2022) Effects of street orientation and tree species thermal comfort within urban canyons in a hot, dry climate. Eco Inform 69:101671
Nazish A, Abbas K, Sattar E (2024) Health impact of urban green spaces: a systematic review of heat-related morbidity and mortality. BMJ Open 14(9):e081632
Norton BA et al (2015) Planning for cooler cities: a framework to prioritise green infrastructure to mitigate high temperatures in urban landscapes. Landsc Urban Plan 134:127–138
Oke TR (1989) The micrometeorology of the urban forest. Philos Trans Roy Soc Lond. B, Biol Sci 324(1223):335–349
Oke TR, Mills G, Christen A, Voogt JA (2017) Urban climates. Cambridge University Press
Pattnaik N et al (2024) Growth and cooling potential of urban trees across different levels of imperviousness. J Environ Manage 361:121242
Pithon JA et al (2021) Grasslands provide diverse opportunities for bird species along an urban-rural gradient. Urban Ecosyst 24(6):1281–1294
Quillet A, Peng C, Garneau M (2010) Toward dynamic global vegetation models for simulating vegetation–climate interactions and feedbacks: recent developments, limitations, and future challenges. Environ Rev 18(NA):333–353
Rahman MA et al (2024) More than a canopy cover metric: influence of canopy quality, water-use strategies and site climate on urban forest cooling potential. Landsc Urban Plan 248:105089
Rahman MA et al (2020a) Tree cooling effects and human thermal comfort under contrasting species and sites. Agric for Meteorol 287:107947
Rahman MA, Moser A, Gold A, Rötzer T, Pauleit S (2018) Vertical air temperature gradients under the shade of two contrasting urban tree species during different types of summer days. Sci Total Environ 633:100–111
Rahman MA, Moser A, Rötzer T, Pauleit S (2019) Comparing the transpirational and shading effects of two contrasting urban tree species. Urban Ecosyst 22:683–697
Rahman MA et al (2020b) Traits of trees for cooling urban heat islands: a meta-analysis. Build Environ 170:106606
R Core Team (2024) R: a language and environment for statistical computing. In: RFFS Computing (Editor), Vienna, Austria
Sanusi R, Johnstone D, May P, Livesley SJ (2016) Street orientation and side of the street greatly influence the microclimatic benefits street trees can provide in summer. J Environ Qual 45(1):167–174
Sanusi R, Johnstone D, May P, Livesley SJ (2017) Microclimate benefits that different street tree species provide to sidewalk pedestrians relate to differences in plant area index. Landsc Urban Plan 157:502–511
Scheuer S et al (2024) A systematic assessment of greening interventions for developing best practices for urban heat mitigation—the case of Huế, Vietnam. Urban Sci 8(2):67
Srivanit M, Jareemit D (2020) Modeling the influences of layouts of residential townhouses and tree-planting patterns on outdoor thermal comfort in Bangkok suburb. J Build Eng 30:101262
Staiger H, Laschewski G, Matzarakis A (2019) Selection of appropriate thermal indices for applications in human biometeorological studies. Atmosphere 10(1):18
Tan H-A et al (2022) Designing and managing biodiverse streetscapes: key lessons from the city of Melbourne. Urban Ecosyst 25(3):733–740
Threlfall CG et al (2017) Increasing biodiversity in urban green spaces through simple vegetation interventions. J Appl Ecol 54(6):1874–1883
Threlfall CG et al (2016) Variation in vegetation structure and composition across urban green space types. Front Ecol Evol 4:66
Tran PT et al (2022) Nature-based solution for mitigation of pedestrians' exposure to airborne particles of traffic origin in a tropical city. Sustain Cities Soc 87:104264

Vos PEJ, Maiheu B, Vankerkom J, Janssen S (2013) Improving local air quality in cities: To tree or not to tree? Environ Pollut 183:113–122

Wang X et al (2023) The influence of vertical canopy structure on the cooling and humidifying urban microclimate during hot summer days. Landsc Urban Plan 238:104841

Wang Y, Ni Z, Peng Y, Xia B (2018) Local variation of outdoor thermal comfort in different urban green spaces in Guangzhou, a subtropical city in South China. Urban For Urban Green 32:99–112

Wu J, Luo K, Wang Y, Wang Z (2021) Urban road greenbelt configuration: the perspective of PM2. 5 removal and air quality regulation. Environ Int 157:106786

Wu Y, Mashhoodi B, Patuano A (2025) Effective street tree and grass designs to cool European neighbourhoods. Urban Clim 61:102376

Xu M, Hong B, Mi J, Yan S (2018) Outdoor thermal comfort in an urban park during winter in cold regions of China. Sustain Cities Soc 43:208–220

Yang Y et al (2018) Simulation on the impacts of the street tree pattern on built summer thermal comfort in cold region of China. Sustain Cities Soc 37:563–580

Zhang T et al (2025) Optimizing pedestrian thermal comfort in urban street canyons for summer and winter: tree planting or low-albedo pavements? Sustain Cities Soc 120:106143

Cooling Strategies and Environmental Indicators: Spatial Assessment of Urban Energy Balance

Hadi RezaeiRad and Saba Sheikhi

Abstract As cities grow, they tend to become significantly warmer than their surrounding areas—a phenomenon known as the Urban Heat Island (UHI) effect. This increased heat not only affects human comfort and health but also raises energy demands, especially in densely built environments. This study investigates how natural elements within cities—such as vegetation, moisture, and green spaces—can help reduce surface temperatures and contribute to urban thermal sustainability. Focusing on Tehran, we analyzed satellite imagery from 2024 to assess how different environmental indicators relate to surface temperature across the city's 22 municipal districts. We used a statistical method called the Band Collection Statistics (BCS) algorithm to examine correlations between land surface temperature and several environmental variables. Among them, four key indicators—vegetation density (NDVI), vegetation moisture (NDMI), surface water presence (NDWI), and urban park density—showed significant relationships with temperature. Results indicate that areas with more water and higher moisture content tend to be cooler than those with vegetation alone. This suggests that water availability and moisture retention may play a more critical role in reducing urban heat than the mere presence of greenery. These findings highlight the importance of integrated blue-green infrastructure strategies that not only increase vegetation but also focus on preserving moisture. The study offers insights for urban planners and policymakers aiming to develop context-specific cooling strategies to enhance livability and environmental resilience in rapidly growing cities like Tehran.

Keywords Cooling strategies · Urban Heat Island (UHI) · Land Surface Temperature (LST) · Blue-green infrastructure · Environmental indicators · Urban Energy Balance (UEB) · Tehran

---

H. RezaeiRad (✉)
Department of Urbanism, Faculty of Art and Architecture, Bu-Ali Sina University, Hamedan, Iran
e-mail: H.rezaeirad@Basu.ac.ir

S. Sheikhi
Department of Urban and Regional Planning, Faculty of Art, Tarbiat Modares University, Tehran, Iran

A. Cheshmehzangi et al. (eds.), *Nature in Cities, Nurturing Cities*, Urban Sustainability,
https://doi.org/10.1007/978-981-95-5487-4_7

1 Introduction

The rapid evolution of urbanization has emerged as one of the significant anthropogenic forces reshaping urban surface energy balance (Rezaeirad 2017). As cities expand both horizontally and vertically, they significantly change how heat is absorbed, stored, and released across their surfaces, thereby creating the Urban Heat Island (UHI) effect (Arnfield 2003)—a phenomenon whereby urban areas become significantly warmer than surrounding rural zones due to the prevalence of heat-retaining surfaces such as asphalt, concrete, and buildings. The consequences of this thermal imbalance extend beyond mere discomfort, encompassing increased energy consumption for cooling, exacerbated air pollution, compromised human health, and amplified climate change impacts at local scales (Santamouris 2020).

Despite occupying a relatively limited portion of the Earth's surface, urban areas are major contributors to global energy consumption and carbon emissions (UN-Habitat 2020). This disproportionate impact arises because cities concentrate populations, buildings, transportation systems, and industrial activity within relatively small geographic areas. These modifications disrupt the natural processes of evapotranspiration, alter albedo rates, and modify airflow patterns, culminating in distinctive urban microclimates (Santamouris et al. 2015). Evapotranspiration is a natural cooling process by which water from soil and plants evaporates into the atmosphere. Albedo refers to how much solar energy is reflected from surfaces—darker surfaces reflect less and absorb more heat. This imbalanced environmental footprint underscores the critical importance of understanding the complex interactions between urban form, natural elements, and thermal energy dynamics.

In recent decades, urban policies have increasingly focused on how morphological development affects environmental sustainability (Alberti 2015). Urban morphology—the shape, density, and structure of cities—plays a key role in how cities handle heat and environmental stress. The thermal energy balance of urban surfaces represents a critical dimension of this inquiry, serving as both an indicator of environmental quality and a determinant of energy consumption patterns. With climate change driving more frequent urban heat events (IPCC 2022), addressing thermal imbalances has become increasingly urgent.

This chapter addresses this imperative by examining the spatial relationship between urban nature components and thermal energy balance. A comprehensive approach is adopted that recognizes the heterogeneity of urban thermal environments and their differential relationships with natural elements across urban space, rather than treating the urban heat island as a uniform phenomenon. The correlation between environmental indicators and surface thermal patterns is investigated using the Band Collection Statistics (BCS) algorithm. The BCS algorithm is a method that helps compare multiple environmental datasets—like vegetation or water maps—with temperature maps to identify statistical relationships.

The research focuses on Tehran, Iran's capital and largest metropolis. The city's distinctive topography creates natural climatic variations that interact with anthropogenic thermal modifications (Sodoudi et al. 2014). This complexity makes Tehran an ideal laboratory for examining the spatial dimensions of urban thermal energy balance and the mitigating potential of natural elements.

By analyzing the correlation between environmental indicators (specifically vegetation density (NDVI), vegetation moisture (NDMI), surface water presence (NDWI) and urban park density) and surface thermal patterns across Tehran's 22 municipal districts, this research aims to elucidate the spatially differentiated relationships between urban nature components and thermal energy balance. By identifying which natural elements exert the strongest mitigating influence on urban heat across different urban contexts, the research provides an empirical foundation for evidence-based green infrastructure strategies. The methodological approach demonstrated here offers a replicable framework for similar analyses in other urban contexts, contributing to the broader scientific understanding of urban thermal dynamics and their modulation through nature-based interventions.

This chapter is structured as follows: it begins with a review of relevant literature on urban thermal dynamics and the role of environmental indicators in moderating heat. The next section describes the data sources and methods, including satellite imagery analysis and the application of the BCS algorithm. The findings section then presents the spatial relationships between key environmental factors and surface temperatures in Tehran. Finally, the chapter concludes with a discussion of implications for urban planning and sustainability, offering practical insights for nature-based solutions and climate adaptation in cities such as Tehran.

2 Literature Review

Urban heat is not a new phenomenon, but understanding its underlying drivers, spatial variability, and mitigation strategies has become increasingly urgent in the context of climate change and rapid urbanization. A growing body of interdisciplinary research—from urban climatology and landscape ecology to remote sensing and urban planning—has sought to explain how cities heat up and stay warm, and how natural elements like vegetation and water bodies can moderate these thermal effects. This section reviews the relevant theoretical and empirical literature, providing the conceptual and methodological foundation for this study. Special attention is given to how thermal energy balance is defined and measured, how satellite-based techniques have advanced our understanding of intra-urban heat patterns, and how environmental indicators such as vegetation indices are used to assess urban thermal dynamics.

2.1 Urban Energy Balance

The conceptual framework for understanding urban thermal energy dynamics has evolved significantly since Howard's (1833) pioneering documentation of London's elevated temperatures compared to its rural surroundings. Howard's early observations marked the beginning of what we now call the Urban Heat Island effect. Contemporary urban climate science conceptualizes the urban thermal environment through the lens of surface energy balance (SEB), expressed as:

$$Q * + Q_F = Q_H + Q_E + \Delta Q_S$$

where Q^* represents net all wave radiation, Q_F denotes anthropogenic heat flux, Q_H indicates sensible heat flux, Q_E signifies latent heat flux, and ΔQ_S represents the net storage heat flux (Rezaeirad 2017).

In conceptual terms, this equation illustrates how heat enters, is stored, and is released at the surface of urban environments. While Q^* and Q_F represent incoming energy—from solar radiation and human activities—the terms on the right side reflect how this energy is transferred to the air (Q_H), used for evaporation (Q_E), or stored in buildings and materials (ΔQ_S). This equation encapsulates how urbanization fundamentally reconfigures energy partitioning by altering each component of the balance. The transformation of natural surfaces to built environments increases net radiation absorption through reduced albedo, diminishes latent heat flux through reduced evapotranspiration, elevates sensible heat flux through increased surface temperatures, augments the storage term through the high thermal capacity of urban materials, and introduces the anthropogenic heat component absent in natural systems (Arnfield 2003; Erell et al. 2011). Oke et al. (2017) elaborate that these modifications create distinctive urban climates characterized by increased sensible heat storage during daytime and prolonged release during nighttime, establishing the classic UHI temporal signature. Table 1 illustrates the bidirectional relationship between urban heat island formation and surface energy balance, highlighting key urban and non-urban characteristics that influence the development of heat islands and the regulation of surface energy (see Table 1).

Recent theoretical advancements have moved beyond this general understanding to examine how specific urban morphological characteristics—including building

Table 1 Urban heat island formation and surface energy balance

Key features contributing to urban heat island formation	Impact on energy balance
Reduction of vegetation cover	Decrease in evapotranspiration
Extensive use of impervious surfaces	Decrease in evapotranspiration
High thermal emissivity of urban materials	Increase in stored heat
Low albedo of urban construction materials	Increase in net radiation
Urban geometry-induced heat retention	Increase in net radiation
Urban geometry-induced reduction in wind speed	Reduction in convective heat transfer
Increase in air pollution levels	Increase in net radiation
Increase in energy consumption	Increase in anthropogenic heat

height-to-width ratios (H/W), sky view factors (SVF), and material properties—modulate the urban energy balance across different spatial scales (Rezaeirad 2017; Wang et al. 2024; Liu et al. 2024).

2.2 Remote Sensing Approaches to Urban Thermal Analysis

Remote sensing has revolutionized the study of urban thermal environments by enabling comprehensive spatial coverage impossible with traditional in-situ measurements (Voogt and Oke 2003). Remote sensing refers to the use of satellite or aerial imagery to collect information about the Earth's surface without physical contact, an essential tool for analyzing temperature patterns across large urban areas.

Satellite thermal infrared (TIR) sensors, including those aboard the Landsat series, MODIS, ASTER, and Sentinel platforms, have become instrumental in quantifying Land Surface Temperature (LST) patterns across urban areas (Sobrino et al. 2008). These sensors detect infrared radiation emitted by surfaces, which is then converted into surface temperature estimates—commonly referred to as Land Surface Temperature (LST). LST is not the same as air temperature, but it is strongly correlated with how much heat a surface absorbs and emits. These technologies have facilitated the transition from conceptualizing the UHI as a city-rural temperature differential to mapping its intraurban spatial heterogeneity (Zhou et al. 2019). In other words, instead of just comparing urban and rural temperatures, researchers now use remote sensing to study detailed heat variations within different parts of the same city.

The methodological approaches for deriving LST from satellite imagery have grown increasingly sophisticated. Early studies relied primarily on the radiative transfer equation with atmospheric correction parameters, while contemporary approaches incorporate enhanced split-window algorithms, mono-window algorithms, and single-channel methods (Jiménez-Muñoz and Sobrino 2003; Jiménez-Muñoz et al. 2014; Wang et al. 2015) to improve accuracy. These methods help translate satellite sensor readings into actual surface temperatures by correcting for atmospheric effects and sensor limitations. The "split-window" and "mono-window" techniques refer to different algorithms for interpreting thermal signals from specific bands.

The temporal dimension of thermal remote sensing has evolved significantly. While earlier studies relied predominantly on single-date imagery, contemporary research increasingly employs multi-temporal analyses to capture seasonal and diurnal variations in thermal patterns (Weng 2009; Zhou et al. 2019). Zhou et al. (2011) demonstrated that such temporal analyses reveal dynamic spatiotemporal characteristics of the UHI that single-date studies might overlook, including seasonal inversions of the UHI effect and variations in the timing of maximum intensity.

2.3 Environmental Indices and Urban Thermal Patterns

The relationship between urban vegetation and thermal conditions has received extensive scholarly attention, with the cooling effect of vegetation well-established through numerous empirical studies (Bowler et al. 2010; Norton et al. 2015). This relationship is commonly quantified using the Normalized Difference Vegetation Index (NDVI), calculated as:

$$\text{NDVI} = (\text{NIR} - \text{Red}) / (\text{NIR} + \text{Red})$$

where NIR represents the near-infrared reflectance and Red the visible red reflectance (Tucker 1979).

NDVI is a widely used index in remote sensing to assess vegetation presence and health. Its values range from -1 to +1, where higher positive values indicate denser, healthier vegetation. Lower or negative values suggest barren land or built-up surfaces. Since vegetation contributes to cooling through shade and evapotranspiration, NDVI is closely linked to land surface temperature in urban areas.

Beyond vegetation density, water bodies and surface moisture content significantly influence urban thermal patterns. The Normalized Difference Water Index (NDWI), formulated as:

$$\text{NDWI} = (\text{NIR} - \text{Green}) / (\text{NIR} + \text{Green})$$

where Green represents the green band reflectance (McFeeters 1996), effectively identifies water bodies and moisture content.

This index highlights areas with surface water or high soil moisture by measuring the difference between how much green light and near-infrared light is reflected. Water absorbs more NIR than vegetation or dry surfaces, making this contrast useful for detecting moist areas.

NDWI values typically range from –1 to +1. Values closer to +1 indicate open water or saturated ground, which tend to have a strong cooling effect in urban areas. In contrast, values near 0 or negative suggest drier surfaces such as built-up areas or bare soil.

Similarly, the Normalized Difference Moisture Index (NDMI), calculated as:

$$\text{NDMI} = (\text{NIR} - \text{SWIR}) / (\text{NIR} + \text{SWIR})$$

where SWIR represents the short-wave infrared reflectance (Jin and Sader 2005), captures vegetation moisture content and is sensitive to drought stress. In this index, values greater than 0.1 indicate a high level of moisture, whereas values approaching -1 represent low moisture.

NDMI helps distinguish between moisture-rich and moisture-stressed vegetation, even when both appear green. In urban climates, this distinction is vital because moisture-stressed vegetation may contribute less to cooling.

2.4 *Spatial Statistical Approaches in Urban Thermal Studies*

Spatial statistics in urban thermal research involve techniques that quantify and model the spatial heterogeneity and distribution of land surface temperature (LST). In this context, spatial heterogeneity refers to how uneven or varied temperature patterns are across different areas of a city. Spatial statistics help researchers not just map these variations but also understand the factors shaping them. These methods include spatial interpolation (e.g., kriging, regression kriging), correlation analysis, and spatial heterogeneity metrics, often integrated with Geographic Information Systems (GIS). Regression kriging is a two-step geostatistical method that first models the relationship between temperature and influencing variables (like vegetation) through regression and then applies kriging to interpolate the residual spatial variation, effectively estimating and spatializing UHI intensity in urban and peri-urban areas with high accuracy and low error (Brabant et al. 2024).

Spatial statistics help identify critical urban heat island patches and their drivers by examining urban form indicators such as building height, roughness, and sky view factor. Sky view factor, for instance, refers to how much of the sky is visible from a certain point at street level. Narrower urban canyons or dense building clusters tend to trap more heat and reduce nighttime cooling. Studies reveal that these factors significantly influence LST distribution, with spatial heterogeneity analysis at the neighborhood or block scale offering insights into heat accumulation and diffusion mechanisms within urban fabrics (Wang et al. 2025).

Band Collection Statistics (BCS) is a tool to provide statistics for the multivariate analysis of a set of raster bands. Raster bands refer to individual layers of data from satellite imagery, such as thermal, near-infrared, or red reflectance. Multivariate analysis using BCS enables an understanding of how different data layers relate to each other. By using the computer correlation matrices option, the correlation matrices, as well as the primary statistical parameters, such as mean, and standard deviation for each layer, are estimated (Snedecor and Cochran 1967).

The correlation matrix gives the value of the correlation coefficients (Alawamy et al. 2020). It refers to the statistical evaluation of relationships among multiple spectral bands or derived indices—such as NDVI, NDWI, and thermal bands—to assess their correlation, variance, and joint influence on land surface temperature and other thermal indicators. Combining thermal bands with other spectral indices like (NDVI) improves the discrimination of urban heat islands from vegetated or water-covered areas, which typically exhibit lower temperatures (Armah et al. 2025).

3 Methodology

3.1 Research Design and Theoretical Basis

The present study is applied in nature and employs a descriptive–analytical approach. The theoretical foundation was developed through a library-based review of existing literature on urban thermal dynamics, remote sensing methods, and environmental indicators affecting land surface temperature (LST). This review enabled the identification of key natural and morphological factors—such as vegetation, surface moisture, and water bodies—that are most relevant to surface energy distribution in urban environments.

3.2 Satellite Imagery and Preprocessing

To capture current thermal and environmental conditions, satellite data were acquired from the NASA EarthData platform, specifically using Landsat 8 imagery. This sensor provides 30-meter resolution multispectral and thermal bands, suitable for urban-scale environmental monitoring. Only cloud-free and atmospherically corrected images from 2024 were selected to ensure data consistency. Preprocessing included geometric correction, atmospheric adjustment, and image clipping to the Tehran metropolitan boundary, using ENVI (v5.6) and ArcGIS Pro (v3.1).

Following the extraction of the satellite image from the Landsat 8 in Operational Land Imager (OLI) sensor, two parallel processes were conducted. Atmospheric and radiometric corrections were performed within the ENVI environment. Subsequently, the Digital Numbers (DN) of thermal bands 10 and 11 were converted into spectral radiance values.

In the Landsat 8 satellite, the calculation of spectral radiance (L_λ) in the sensor is expressed in units of ($\mathrm{W/(m^2{\cdot}sr{\cdot}\mu m)}$), is defined by the following equation (Yuan and Bauer 2007):

$$L_\lambda = \text{Grescale} * \text{QCAL} + \text{Brescale}$$

Moreover, the following equation may also be utilized:

$$L_\lambda = (L_{\max} - L_{\min}/\text{QCAL}_{\max} - \text{QCAL}_{\min}) * (\text{QCAL} - \text{QCAL}_{\min}) + L_{\min}$$

In the above equations, QCAL refers to the quantized calibrated pixel value in Digital Numbers (DN). Grescale represents the factor used to scale up the digital numbers (DN) of a specific band, and Brescale denotes the factor used to adjust the scaling of a specific band's digital numbers (DN). Additionally, $\text{QCAL}_{\max}$ indicates the maximum quantized calibrated pixel value, and $\text{QCAL}_{\min}$ refers to the minimum quantized calibrated pixel value. L_{min} is the spectral radiance corresponding to $QCAL_{min}$, while L_{max} is the spectral radiance associated with $QCAL_{max}$.

In the next step, the spectral radiance layer was converted into Brightness Temperature (BT) as recorded by the satellite sensor. Subsequently, the Brightness Temperature was converted from Kelvin to Celsius using the following equation:

$$BT_c = BT_k - 273.15$$

3.3 Derivation of Environmental Indicators

From the satellite images, a series of spectral indices were calculated to represent key natural features influencing urban temperature. These include the Normalized Difference Vegetation Index (NDVI) for vegetation density, the Normalized Difference Water Index (NDWI) for surface water and soil moisture, and the Normalized Difference Moisture Index (NDMI) for vegetation canopy moisture. In addition, urban park density was computed using open space data to capture human-planned greenery. All indices were standardized to 30-meter resolution to match the spatial scale of the thermal imagery.

3.4 Land Surface Temperature (LST) Estimation

Land Surface Temperature (LST) was derived from Landsat 8's thermal infrared band through a three-step process: radiometric calibration, conversion to brightness temperature, and correction for surface emissivity. Surface emissivity values were estimated by first classifying the land cover using the raw satellite imagery, followed by assigning emissivity values to each land cover class. The calibrated and corrected thermal data were then converted into Celsius degrees to generate a continuous spatial map of surface temperature across Tehran. This LST layer served as the basis for spatial correlation with selected environmental indicators.

3.5 Spatial Framework and Analytical Grid

To conduct a uniform analysis, a regular spatial grid of $1{,}000 \times 1{,}000\text{-m}^2$ cells was applied across the entire metropolitan area. This resolution corresponds to the typical neighborhood scale used in urban planning, ensuring that comparisons of environmental and thermal conditions are meaningful and consistent across districts. Within each cell, the average value of each environmental variable and LST was calculated.

3.6 Correlation Analysis Using Band Collection Statistics (BCS)

The Band Collection Statistics (BCS) algorithm in ENVI was used to quantify the statistical relationships between environmental indicators and LST. This tool computes correlation matrices, providing Pearson correlation coefficients along with basic statistical descriptors (mean, standard deviation, etc.) for each raster layer. The correlation values help assess both the strength and direction (positive or negative) of relationships between temperature and natural components across space.

3.7 Spatial Interpretation and Visualization

Finally, the correlation outputs were imported into GIS for mapping and visualization purposes. Thematic layers were generated for LST and each environmental variable, along with correlation intensity heatmaps, to enable spatial representation of the analyzed relationships.

3.8 Study Area: Tehran Metropolis

Tehran, the capital and largest city of Iran, serves as the spatial focus of this study (see Fig. 1). Geographically, it is situated on the southern slopes of the Alborz Mountain range, between 52°37′31″ and 52°44′25″ E in longitude and between 36°30′14″ and 36°35′30″ N in latitude (Firozjaei et al. 2019). The city spans an area of approximately 800 square kilometers and forms the core of the wider Tehran Metropolitan Region. According to the World Population Review (2025), Tehran hosts over 9.7 million residents within its municipal boundaries, while the broader urban agglomeration exceeds 15 million inhabitants, making it one of the most densely populated megacities in West Asia (Rabiei and Kazemi 2016).

Topographically, Tehran is characterized by a pronounced elevation gradient that spans from over 1,800 meters above sea level in the north to approximately 1,000 meters in the southern districts (Zargari et al. 2025). This north-to-south slope significantly influences the city's microclimates, wind patterns, and surface temperature variability. The northern districts, adjacent to the Alborz Mountains, are generally cooler, greener, and subject to orographic airflow, while the southern and southeastern zones are flatter, more densely built, and typically experience higher surface temperatures. The topographic contrast also creates variation in air pollution accumulation, precipitation rates, and vegetation distribution, all of which are relevant to understanding the city's surface energy balance.

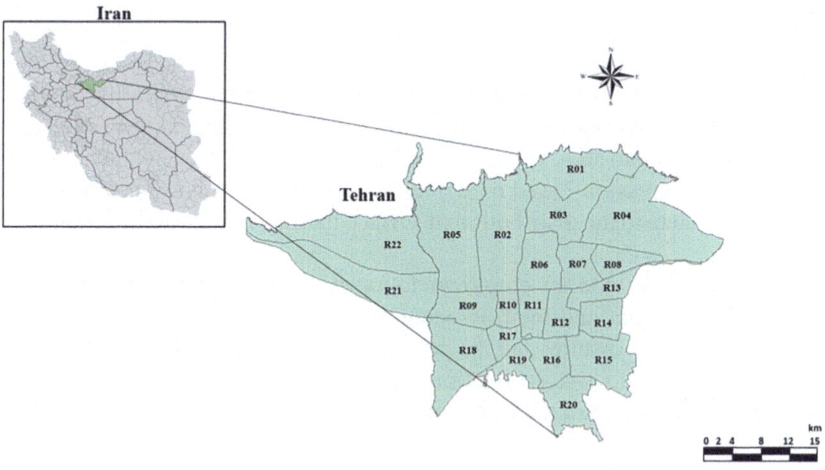

Fig. 1 Research area location

From a morphological perspective, Tehran displays highly heterogeneous urban structure. The central and southern districts are characterized by high-density residential and commercial development, limited vegetation cover, and compact street grids, often associated with elevated urban heat. In contrast, the northern and northwestern districts include a mix of high-income residential neighborhoods, hilly terrain, and more substantial green cover, including parks, gardens, and tree-lined streets. Urban expansion over recent decades has led to significant peripheral growth toward the west and southwest, creating a mix of formal and informal urban fabrics with varying access to infrastructure and green space.

Administratively, Tehran is divided into 22 municipal districts, each exhibiting distinct environmental and socio-economic characteristics. For instance, District 1 in the northeast includes affluent hillside neighborhoods with cooler microclimates and abundant vegetation, while Districts 17, 18, and 19 in the south are more industrialized and suffer from a combination of urban heat accumulation, poor ventilation, and lower vegetation indices. The presence of major transportation corridors, large parks, and urban voids such as dried riverbeds and abandoned industrial zones also contribute to the spatial variation in environmental performance.

In sum, Tehran's combination of steep elevation changes, diverse land use types, and pronounced socio-environmental disparities makes it an exemplary case for analyzing urban surface thermal energy dynamics. The spatial structure of the city provides both the complexity and the contrast needed to examine how different nature-based elements modulate surface temperature patterns under real-world urban conditions.

4 Findings

This section presents the key analytical results derived from the spatial and statistical assessments of land surface temperature (LST) and environmental indicators across the Tehran metropolitan area. By integrating remote sensing data, spatial correlation algorithms, and regression-based modeling, the study reveals how specific environmental indices relate to the distribution of surface thermal energy at the urban district scale.

These findings provide an empirical foundation for understanding the spatial dynamics of urban heat and their association with natural elements such as vegetation cover, surface moisture, and the presence of urban parks. The results outlined below offer both quantitative correlations and spatial visualizations that inform the broader discussion on urban thermal sustainability.

4.1 Spatial Distribution of Land Surface Temperature (LST)

The final land surface temperature map for the Tehran metropolitan area in August 2024 (see Fig. 2) reveals considerable spatial variation. The minimum temperature is estimated at 24.51 °C, while the maximum reaches 43.11 °C, reflecting a substantial urban thermal gradient likely influenced by varying environmental and morphological features.

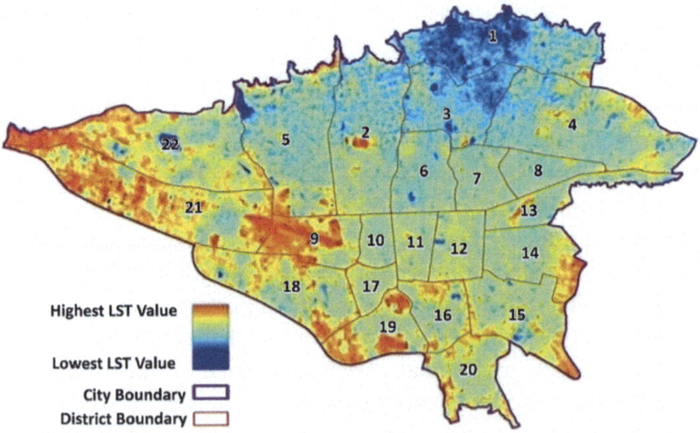

Fig. 2 Land surface temperature map of Tehran metropolitan area, August 2024

Table 2 Spatial correlation assessment of environmental indices with the thermal energy layer of the Tehran metropolitan

Indices	The correlation with LST in the 22 districts of Tehran in 2024 (%)	Type of relationship
NDMI	60	Inverse
NDVI	49	Inverse
NDWI	36	Inverse
Number of parks	28	Inverse

4.2 Spatial Correlation between Environmental Indices and Surface Thermal Patterns

To understand the environmental drivers behind LST variation, spatial correlation analysis was conducted using the Band Collection Statistics (BCS) algorithm. This method estimated the relationship between each environmental index and the surface thermal energy layer across Tehran's 22 districts. Among all the environmental factors, four indices—NDMI, NDVI, NDWI, and the number of urban parks and green spaces— with a correlation above 25% were selected for detailed analysis (see Table 2).

These results indicate a clear inverse relationship: higher environmental index values are generally associated with lower surface temperatures, emphasizing the cooling role of moisture, vegetation, and green space distribution.

4.3 Visual Interpretation of Spatial Patterns in Key Environmental Indices

Figure 2 presents the spatial distribution maps of NDMI, NDWI, and NDVI across Tehran. These maps reveal that NDMI and NDVI values are notably higher in northern districts, where vegetation and soil moisture are more abundant. NDWI values, indicating surface water presence, are sporadically distributed but tend to align with rivers, parks, and other blue-green infrastructures. The spatial juxtaposition of these indices reinforces their combined influence in regulating LST (see Fig. 3).

4.4 Integration of Environmental Indices at the Metropolitan Scale

Using a pixel size of 1000 × 1000 meters, the spatial combination of the selected indices was carried out to estimate their joint influence on urban thermal energy,

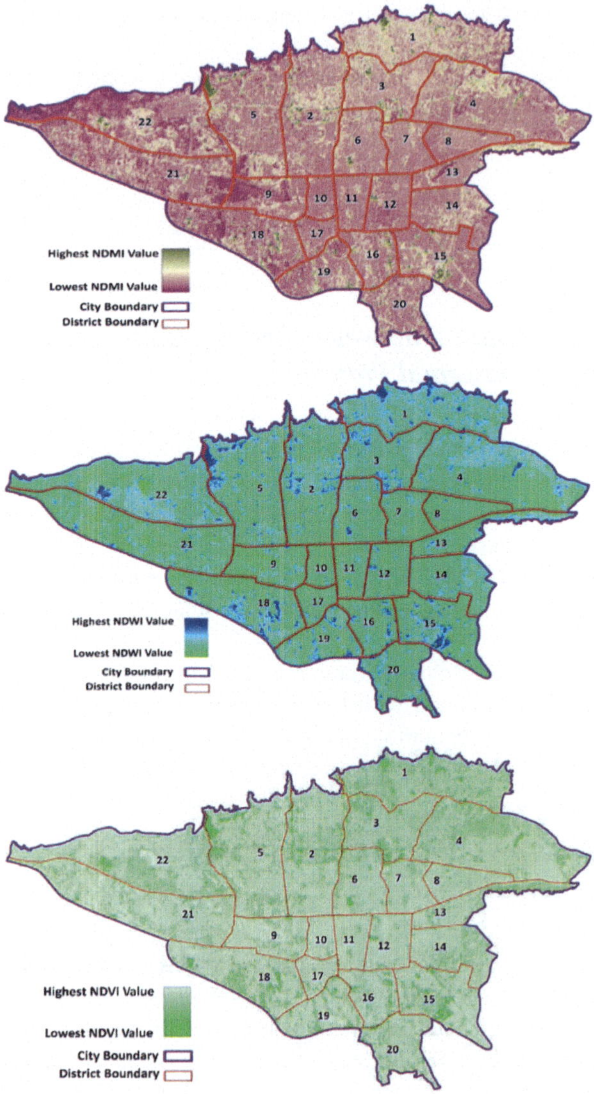

Fig. 3 Spatial distribution maps of the indices: (top) NDMI, (middle) NDWI, and (bottom) NDVI in Tehran

considering the area of neighboring units for the entire Tehran metropolitan (see Fig. 4).

The central, eastern, and western axes of the city exhibited the lowest composite environmental index values—aligning with areas of high surface temperature—while the north maintained the highest environmental index concentrations and cooler conditions.

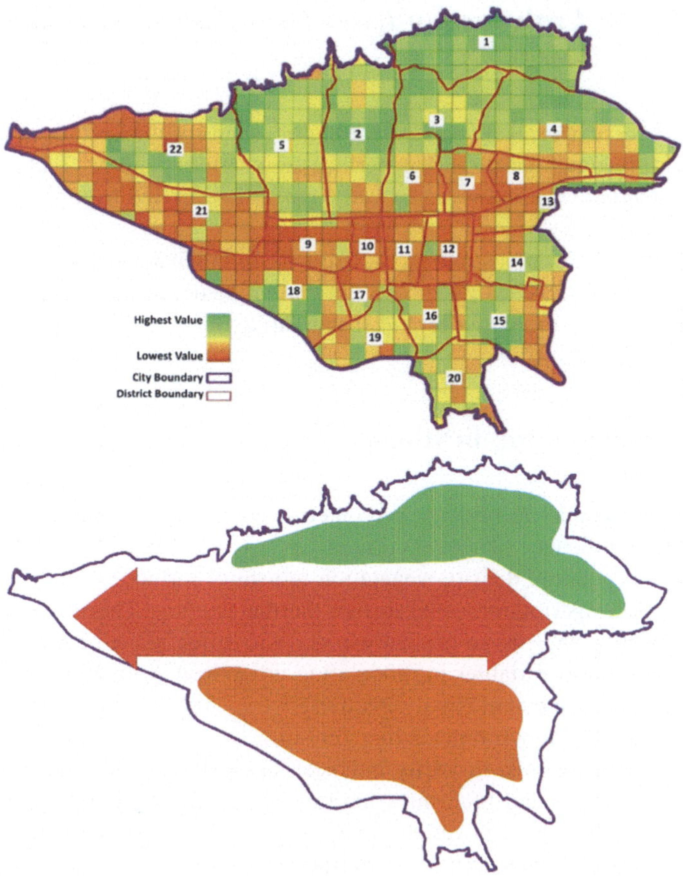

Fig. 4 Spatial estimation of environmental indices in the thermal energy balance of Tehran metropolitan area (top) and Schematic graph of the correlation of environmental components in the thermal energy balance of Tehran metropolis (bottom).

A schematic representation in Fig. 4 further illustrates how these indices jointly contribute to the thermal energy balance of the urban surface, emphasizing the cumulative cooling potential of combined green and blue elements.

4.5 *Statistical Evaluation of Index Contributions to Urban Thermal Energy*

Statistical regression analysis was performed to quantify the specific contribution of each environmental index to urban surface temperature patterns. Table 3 summarizes the unstandardized and standardized coefficients derived from the model (see Table 3).

These coefficients demonstrate the statistically significant and relatively balanced contribution of NDWI, NDMI, and NDVI in mitigating urban surface heat, with park distribution playing a more modest but still notable role.

5 Discussion and implications

Building on the empirical findings presented in the previous section, this chapter interprets the significance of spatial correlations between key environmental indicators—namely vegetation density, vegetation moisture, surface water presence, and urban green space distribution—and surface thermal patterns. This discussion moves beyond descriptive analysis to critically assess the differentiated roles and relative contributions of these variables in modulating urban heat, with particular attention to spatial variability and contextual sensitivity.

In semi-arid urban environments like Tehran, where water scarcity, rapid urbanization, and socioeconomic disparity intersect, understanding the nuanced thermal performance of green and blue infrastructure is central to advancing sustainable urban planning. The observed variations in correlation strength across districts, and among indicators, challenge uniform assumptions about the cooling efficacy of urban greenery and call for more spatially tailored, performance-oriented approaches.

This section not only elucidates these spatially heterogeneous relationships but also explores their implications for planning practice, environmental equity, and thermal resilience policy. Finally, the discussion reflects on the methodological

Table 3 Statistical analysis of the coefficient values of environmental indices influencing urban thermal energy.

Coefficients[a]					
Model	Unstandardized coefficients		Standardized coefficients		
	B	Std. Error	Beta	t	Sig.
(Constant)	−1.08	0.000		−4.786E7	0.000
NDWI	0.347	0.000	0.341	6.299E7	0.000
NDMI	0.320	0.000	0.320	9.985E7	0.000
Number of parks	0.131	0.000	0.131	9.501E7	0.000
NDVI	0.287	0.000	0.342	4.754E7	0.000

contributions and limitations of the current approach, offering insights for future research and practical application. In doing so, it aims to bridge the gap between scientific analysis and actionable urban policy, aligning remote sensing-based thermal studies with the real-world imperatives of climate adaptation and urban sustainability.

5.1 *Quantifying Environmental Contributions to Surface Thermal Patterns*

Based on the findings of the study, considering the spatial pattern of surface heat and environmental indices in the Tehran metropolitan area, and spatial regression calculations in a GIS environment, along with statistical calculations based on the coefficients of influential environmental dimensions on the dependent variable in Table 3, the equation describing the relationship between independent variables influencing the energy balance of urban surfaces in Tehran can be defined as follows:

$$
\begin{aligned}
A &= -1.08 + (0.347a_1 + 0.322a_2 + 0.131a_3 + 0.287a_4) \\
A &= \text{Environmental factor} \\
a_1 &= \text{NDWI index} \\
a_2 &= \text{NDMI index} \\
a_3 &= \text{Number of parks and green spaces} \\
a_4 &= \text{NDVI index}
\end{aligned}
$$

The analysis of the spatial correlation between environmental indicators and surface thermal patterns across Tehran's 22 districts reveals significant heterogeneity in these relationships, defying simplistic generalizations about how natural elements cool urban environments. The differential correlation strengths observed suggest that the thermal impact of urban nature components varies substantially based on their spatial context, composition, and configuration.

5.2 *Evaluating the Role of Water and Vegetation in Urban Cooling*

The finding that NDWI exhibits the strongest overall correlation with surface thermal patterns (0.34) underscores the exceptional cooling efficiency of water in Tehran's semi-arid context. This aligns with previous studies by Guha et al. (2020) and Sun and Chen (2012), who similarly identified water's superior cooling capacity in arid urban environments. The physiological basis for this phenomenon lies in water's high specific heat capacity and the substantial energy consumption through evaporation. In Tehran's water-limited environment, even small increases in surface moisture produce disproportionate cooling effects, explaining the strong NDWI-LST relationship observed.

The significant correlation between NDMI and surface thermal patterns (0.32) further reinforces the critical role of moisture in urban thermal regulation. Unlike NDWI, which primarily captures water bodies and saturated surfaces, NDMI detects moisture content within vegetation and soil (Gao 1996). Its strong correlation suggests that in Tehran's semi-arid climate, vegetation moisture content—not merely vegetation presence—constitutes a critical determinant of cooling efficacy. This finding resonates with Chen et al.'s (2006) observation that vegetation moisture status significantly modulates the cooling effect of urban greenery, with drought-stressed vegetation providing substantially reduced thermal benefits.

The comparatively weaker correlation observed between NDVI and surface thermal patterns (0.28), while still significant, challenges the conventional prioritization of vegetation quantity in urban cooling strategies. This finding does not contradict the established cooling effect of vegetation documented in numerous studies but suggests that in Tehran's context, vegetation quality—particularly its moisture status—may prove more thermally significant than mere quantity. The weaker NDVI correlation likely reflects the presence of drought-stressed vegetation in many areas of Tehran, which despite registering as "green" in NDVI calculations, provides limited cooling through evapotranspiration due to moisture conservation strategies (stomatal closure) adopted under water stress.

The weakest correlation observed with urban park density (0.13) appears counterintuitive given the documented cooling effects of urban parks. However, this finding may reflect limitations in the park density metric, which quantifies park area without accounting for critical qualitative factors such as vegetation type, irrigation status, design configuration, and maintenance quality—all factors that significantly influence cooling performance. Additionally, this weaker correlation may indicate that Tehran's parks, especially in water-stressed southern districts, often contain significant non-vegetated areas (plazas, pathways, sports facilities) that contribute minimally to cooling despite being classified as park space.

5.3 Spatial Variation and District-Level Disparities

The spatial analysis reveals notable district-level variations in correlation strengths, with several patterns emerging. First, correlations between all environmental indicators and surface temperature strengthen along the north-south gradient, generally exhibiting stronger negative correlations in northern districts. This pattern likely reflects both the greater abundance of well-maintained vegetation in these affluent areas and the steeper topography that enhances air circulation and cooling efficacy. Second, correlation strengths show greater spatial consistency for NDWI compared to NDVI and NDMI, suggesting that water's cooling effect remains relatively stable across different urban contexts, while vegetation's cooling impact varies more substantially with urban configuration and maintenance practices.

These spatially differentiated relationships highlight the limitations of non-spatial analytical approaches that calculate city-wide correlation coefficients and develop universal prescriptions for urban cooling. The significant intra-urban variation in correlation strengths observed in this study underscores the necessity of spatially explicit analyses that can capture these heterogeneous relationships and inform targeted, context-sensitive cooling interventions.

5.4 Cooling Strategies for Climate-Responsive Urban Planning

The findings of this spatial analysis yield important implications for urban planning practices aimed at enhancing thermal sustainability in Tehran and similar semi-arid urban contexts. First, the strong cooling association of water and moisture indicators (NDWI, NDMI) relative to vegetation density alone (NDVI) suggests that urban cooling strategies should prioritize integrated blue-green approaches rather than focusing exclusively on expanding green coverage. Specifically, planning efforts should emphasize moisture-preserving designs such as rain gardens, bioswales, and detention basins that capture and retain precipitation, maximizing its cooling potential through prolonged evaporation. The implementation of permeable pavements, in the heavily sealed southern districts, represents another important strategy for enhancing moisture retention and increasing latent heat flux.

Second, the relatively modest correlation with park density highlights the need to supplement quantitative park provision targets with qualitative standards addressing park design, composition, and maintenance. The current Tehran Comprehensive Plan establishes minimum per capita green space requirements (7-12 m^2/person depending on district) but lacks specific guidelines for the thermal performance of these spaces. Incorporating thermal design principles—such as optimal tree placement, species selection for maximum evapotranspiration, irrigation infrastructure, and the integration of water features—could significantly enhance the cooling contribution of both existing and new park spaces.

Third, the spatial heterogeneity in environmental indicator-thermal relationships across Tehran's districts necessitates geographically differentiated cooling strategies rather than uniform city-wide approaches. The stronger correlation between vegetation indicators and surface temperature in northern districts suggests these areas would benefit most from conventional greening approaches, while the weaker correlations in southern districts indicate that more intensive interventions—potentially emphasizing blue infrastructure and high-efficiency cooling vegetation—may be required to achieve comparable thermal benefits in these areas.

Fourth, the findings have significant environmental justice implications given the alignment between thermal vulnerability and socioeconomic disadvantage in Tehran. The weaker cooling associations in lower-income southern districts, coupled with their higher baseline temperatures and more limited green infrastructure, creates a compound disadvantage requiring prioritized intervention. Targeted investments in high-performance cooling infrastructure in these vulnerable districts would address both environmental and social equity objectives.

Fifth, the observed relationships highlight the importance of water management and conservation in Tehran's urban cooling strategy. The city's limited water resources—constrained by climate change and population growth—necessitate efficient approaches that maximize cooling benefits per unit water input. This might include prioritizing drought-resistant but high-transpiration species, implementing precision irrigation systems, promoting rainwater harvesting for landscape irrigation, and recycling greywater for urban vegetation.

Finally, the significant influence of all studied environmental indicators on surface thermal patterns emphasizes the importance of integrating thermal considerations into broader urban development regulations. Beyond dedicated green and blue infrastructure initiatives, thermal performance criteria should inform building codes, zoning regulations, and development approval processes to ensure that new urban development enhances rather than deteriorates the city's thermal environment. This might include requirements for minimum vegetation coverage, green roofs on new buildings, permeable surface minimums, and building orientation optimization for passive cooling.

5.5 Methodological Considerations and Future Research Directions

Without substantial intervention, the combination of rising baseline temperatures and the urban heat island effect could create thermal conditions that exceed human adaptation thresholds in many parts of the city during summer, with disproportionate impacts on vulnerable populations.

This study's application of the Band Collection Statistics (BCS) algorithm to analyze the spatial correlation between environmental indicators and surface thermal patterns represents a methodological advancement in urban thermal research, offering advantages over conventional global correlation approaches. Nevertheless, several methodological considerations and limitations warrant discussion to contextualize the findings and inform future research.

The cross-sectional nature of the analysis, based on a single-date satellite image, represents another methodological limitation. Urban thermal patterns and their relationships with environmental indicators exhibit significant temporal variability across diurnal and seasonal cycles. The summer daytime image analyzed likely captures maximum UHI intensity but may not represent relationships during other seasons or at night when thermal dynamics differ substantially. Future research should incorporate multi-temporal analyses spanning different seasons and times of day to develop a more comprehensive understanding of how environmental indicator-thermal relationships evolve temporally.

While the study identifies significant correlations between environmental indicators and surface thermal patterns, correlation does not necessarily imply causation. The observed relationships may be influenced by confounding variables not included in the analysis, such as building density, anthropogenic heat emissions, material properties, and air circulation patterns. Future research should employ multivariate analytical approaches, such as structural equation modeling or partial least squares regression, to disentangle these complex interrelationships and isolate the specific contribution of each environmental indicator to thermal patterns.

The study's focus on surface temperature, while methodologically necessary for remote sensing analysis, represents an indirect proxy for actual thermal comfort experienced by urban residents, which depends on air temperature, humidity, wind speed, and radiative conditions. The relationship between surface and air temperatures varies across different urban contexts and temporal conditions. Future research should complement satellite-based surface temperature analyses with ground-based measurements of air temperature and thermal comfort indices to establish more direct links between environmental indicators and human thermal experience.

Despite these limitations, the study's methodological approach offers a valuable template for spatially explicit analysis of urban thermal patterns that could be refined and expanded in future research. The integration of advanced spatial statistical techniques with remote sensing data represents a promising direction for understanding and addressing urban thermal challenges in the context of rapid urbanization and climate change.

6 Concluding Remarks

This chapter has examined the spatial relationship between urban nature components and surface thermal energy balance in Tehran metropolis, employing the Band Collection Statistics algorithm to quantify the correlation between environmental indicators and thermal patterns across the city's 22 municipal districts. The analysis reveals several key findings with important implications for urban thermal management.

First, water-related indicators—specifically NDWI and NDMI—demonstrate stronger correlations with surface thermal patterns than vegetation density alone, highlighting the critical role of moisture in regulating urban thermal environments, particularly in semi-arid contexts like Tehran. These findings challenge conventional urban cooling approaches that prioritize vegetation quantity over moisture retention and suggest that integrated blue-green infrastructure strategies may offer superior thermal benefits in water-limited urban environments.

Second, the analysis reveals substantial spatial heterogeneity in environmental indicator-thermal relationships across Tehran's diverse urban landscape. These spatially differentiated correlations underscore the limitations of non-spatial analytical approaches and highlight the necessity of context-sensitive cooling strategies tailored to local urban conditions. The stronger cooling associations observed in Tehran's northern districts suggest conventional greening approaches prove more effective in these areas, while the weaker associations in southern districts indicate more intensive interventions may be required to achieve comparable thermal benefits.

Third, the modest correlation between park density and surface thermal patterns suggests that the cooling contribution of urban parks depends not merely on their spatial extent but on qualitative factors such as design, composition, maintenance, and integration with water systems. This finding highlights the importance of supplementing quantitative green space targets with qualitative standards addressing the thermal performance of urban nature components.

From a methodological perspective, this study demonstrates the value of spatially explicit analytical approaches in urban thermal research. The BCS algorithm's capacity to capture intra-urban variation in environmental indicator-thermal relationships provides insights unobtainable through conventional global correlation analyses, offering a more comprehensive understanding of how urban nature components influence thermal conditions across different urban contexts.

These findings hold significant implications for urban planning and design practice in Tehran and similar semi-arid cities. They suggest that enhancing urban thermal sustainability requires moving beyond simplistic greening targets toward integrated approaches that:

1. Prioritize moisture retention and management through permeable surfaces, water-sensitive urban design, and efficient irrigation systems
2. Develop context-sensitive cooling strategies tailored to local urban conditions rather than implementing uniform city-wide approaches
3. Address qualitative aspects of urban nature components through thermal performance standards for green spaces
4. Target interventions to address environmental justice concerns, prioritizing cooling infrastructure in vulnerable districts
5. Integrate thermal considerations into broader urban planning frameworks, including building codes, zoning regulations, and development approval processes.

Future research should build upon these findings by incorporating multi-temporal analyses, expanding the range of environmental indicators examined, integrating ground-based thermal measurements, and employing simulation approaches to project the impact of potential interventions. Additionally, research exploring the intersection of urban thermal patterns with social vulnerability factors would enhance understanding of the environmental justice dimensions of urban heat and inform more equitable cooling strategies.

In conclusion, this study underscores the critical importance of urban nature components in regulating thermal energy balance and mitigating urban heat. By revealing the spatially differentiated relationships between these components and surface thermal patterns, it provides valuable insights into developing targeted, context-sensitive interventions that enhance urban thermal sustainability through nature-based solutions. As cities worldwide confront the dual challenges of rapid urbanization and climate change, such evidence-based approaches to urban thermal management become increasingly essential for creating urban environments that remain livable, sustainable, and resilient in the face of rising temperatures.

References

Alawamy JS, Balasundram SK, Mohd Hanif AH, Boon Sung CT (2020) Detecting and analyzing land use and land cover changes in the region of Al-Jabal Al-Akhdar, Libya using time-series landsat data from 1985 to 2017. Sustainability 12(11):4490

Alberti M (2015) Eco-evolutionary dynamics in an urbanizing planet. Trends Ecol Evol 30(2):114–126

Armah RN, Ning ZH, Twumasi YA, Osei JD, Masasi B, Anokye M, Loh PM (2025) Mapping the spatial distribution of urban heat island in Scotlandville in the Louisiana state of USA using satellite remote sensing. The Int Arch Photogrammetry, Remote Sens Spatial Inf Sci 48:9–14

Arnfield AJ (2003) Two decades of urban climate research: a review of turbulence, exchanges of energy and water, and the urban heat island. Int J Climatol: A J Roy Meteorol Soc 23(1):1–26

Bowler DE, Buyung-Ali L, Knight TM, Pullin AS (2010) Urban greening to cool towns and cities: a systematic review of the empirical evidence. Landsc Urban Planning 97(3):147–155

Brabant C, Dubreuil V, Dufour S (2024) Evaluation of spatial interpolation techniques for urban heat island monitoring in small and medium sized cities. Front Built Environ 10:1455047

Chen XL, Zhao HM, Li PX, Yin ZY (2006) Remote sensing image-based analysis of the relationship between urban heat island and land use/cover changes. Remote Sens Environ 104(2):133–146

Erell E, Pearlmutter D, Williamson T (2011) Urban microclimate: designing the spaces between buildings. Routledge

Firozjaei MK, Alavipanah SK, Liu H, Sedighi A, Mijani N, Kiavarz M, Weng Q (2019) A PCA–OLS model for assessing the impact of surface biophysical parameters on land surface temperature variations. Remote Sens 11(18):2094

Gao BC (1996) NDWI—a normalized difference water index for remote sensing of vegetation liquid water from space. Remote Sens Environ 58(3):257–266

Guha S, Govil H, Diwan P (2020) Analytical study on the relationship between land surface temperature and land use/land cover indices. Ann GIS 26(2):201–216

Howard L (1833) The climate of London, deduced from meteorological observations, made at different places in the neighbourhood of the metropolis, vol 1–3. Harvey and Darton

IPCC (2022) Climate change 2022: impacts, adaptation and vulnerability. contribution of working group II to the sixth assessment report of the intergovernmental panel on climate change. Cambridge University Press.

Jiménez-Muñoz JC, Sobrino JA (2003) A generalized single-channel method for retrieving land surface temperature from remote sensing data. J Geophys Res: Atmos 108(D22)

Jiménez-Muñoz JC, Sobrino JA, Skoković D, Mattar C, Cristóbal J (2014) Land surface temperature retrieval methods from Landsat-8 thermal infrared sensor data. IEEE Geosci Remote Sens Lett 11(10):1840–1843

Jin S, Sader SA (2005) Comparison of time series tasseled cap wetness and the normalized difference moisture index in detecting forest disturbances. Remote Sens Environ 94(3):364–372

Liu B, Liu Y, Cho S, Chow DHC (2024) Urban morphology indicators and solar radiation acquisition: 2011–2022 review. Renew Sustain Energy Rev 199:114548

McFeeters SK (1996) The use of the normalized difference water index (NDWI) in the delineation of open water features. Int J Remote Sens 17(7):1425–1432

Norton BA, Coutts AM, Livesley SJ, Harris RJ, Hunter AM, Williams NS (2015) Planning for cooler cities: a framework to prioritise green infrastructure to mitigate high temperatures in urban landscapes. Landsc Urban Planning 134:127–138

Oke TR, Mills G, Christen A, Voogt JA (2017) Urban climates. Cambridge University Press

Rabiei-Dastjerdi H, Kazemi M (2016) Tehran: old and emerging spatial divides. In: Arefian F, Moeini S (eds) Urban change in Iran: stories of rooted histories and ever-accelerating developments. Springer International Publishing, Cham, pp 171–186
RezaeiRad H (2017) Analysis of physical planning effects on energy consumption balance in Tehran Metropolitan Sub Regions, Dissertation for the Degree of Doctor of Philosophy (PhD), Tarbiat Modares, Tehran
Santamouris M (2020) Recent progress on urban overheating and heat island research. Integrated assessment of the energy, environmental, vulnerability and health impact. Synergies with the global climate change. Energy Build 207:109482
Santamouris M, Cartalis C, Synnefa A, Kolokotsa D (2015) On the impact of urban heat island and global warming on the power demand and electricity consumption of buildings—a review. Energy Build 98:119–124
Snedecor GW, Cochran WA (1967) Statistical methods, 6th edn. Iowa State University Press, Ames, IA
Sobrino JA, Jiménez-Muñoz JC, Sòria G, Romaguera M, Guanter L, Moreno J, Plaza A, Martínez P (2008) Land surface emissivity retrieval from different VNIR and TIR sensors. IEEE Trans Geosci Remote Sens 46(2):316–327
Sodoudi S, Shahmohamadi P, Vollack K, Cubasch U, Che-Ani AI (2014) Mitigating the urban heat island effect in megacity Tehran. Adv Meteorol 2014(1):547974
Sun R, Chen L (2012) How can urban water bodies be designed for climate adaptation? Landsc Urban Planning 105(1–2):27–33
Tucker CJ (1979) Red and photographic infrared linear combinations for monitoring vegetation. Remote Sens Environ 8(2):127–150
UN-Habitat (2020) World cities report 2020: the value of sustainable urbanization. United Nations Human Settlements Programme
Voogt JA, Oke TR (2003) Thermal remote sensing of urban climates. Remote Sens Environ 86(3):370–384
Wang F, Qin Z, Song C, Tu L, Karnieli A, Zhao S (2015) An improved mono-window algorithm for land surface temperature retrieval from Landsat 8 thermal infrared sensor data. Remote Sens 7(4):4268–4289
Wang J, Meng F, Jing T, Qi L (2025) Correlation analysis of the spatial characteristics and influencing factors of the block-scale thermal environment: a case study in Jinan, China. Polish J Environ Stud
Wang Y, Pan W, Liao Z (2024) Impact of urban morphology on high-density commercial block energy consumption in severe cold regions. Sustainability 16(13):5795
Weng Q (2009) Thermal infrared remote sensing for urban climate and environmental studies: methods, applications, and trends. ISPRS J Photogrammetry Remote Sens 64(4):335–344
Yuan F, Bauer ME (2007) Comparison of impervious surface area and normalized difference vegetation index as indicators of surface urban heat island effects in Landsat imagery. Remote Sens Environ 106(3):375–386
Zargari M, Mofidi A, Entezari A, Baaghideh M (2025) Climatology of Tehran surface heat Island: a satellite-based spatial analysis. Sci Rep 15(1):10567
Zhou D, Xiao J, Bonafoni S, Berger C, Deilami K, Zhou Y, Frolking S, Yao R, Qiao Z, Sobrino J (2019) Satellite remote sensing of surface urban heat islands: progress, challenges, and perspectives. Remote Sens 11(1):48
Zhou W, Huang G, Cadenasso M (2011) Does spatial configuration matter? Understanding the effects of land cover pattern on land surface temperature in urban landscapes. Landsc Urban Planning 102(1):54–63

Social Belonging and Placemaking

Transformative Nature-Driven Placemaking for Sheikhupura, Pakistan

Rafia Gulzar, Saima Gulzar, and Saad Mujahid

Abstract The current study aimed to explore the transformative nature-driven placemaking for the haphazardly grown city of Sheikhupura, Pakistan. Like many other urbanizing cities, Sheikhupura is facing challenges related to unplanned expansion, environmental degradation, and the erosion of public urban spaces. Unplanned growth and economic instability are the major contributing factors for spatial fragmentation and decayed built environment with lack of urban public spaces. By harnessing the power of Nature-based Solutions (NbS), this approach aims to restore ecological balance, promote social inclusion, and improve the overall sustainability of public urban spaces. The revitalization of existing green areas in Sheikhupura may significantly contribute to mitigating urban decay. The existing parks, urban forests, and public spaces often suffer from neglect, underutilization, or lack of biodiversity. Using a mixed-methods approach, combining literature mapping, survey research, and stakeholder interviews, this study investigated how revitalization of existing green areas would drive green future. Through targeted interventions, such as enhancing green corridors, improving the quality of green infrastructure, and increasing biodiversity, these spaces can be transformed into vibrant community hubs. Simultaneously, planning for future green spaces is vital to ensuring the long-term sustainability and liveability of Sheikhupura city. The creation of a connected network of green spaces through urban design, such as public parks, community gardens, and green pathways, would not only improve air quality and biodiversity, but also foster a sense of belonging and community engagement. These spaces should be designed with local input to ensure they meet the needs of all residents, particularly marginalized communities. Community participatory approaches are key to the success of this strategy. Involving residents in the design/planning, development, and maintenance of green spaces fosters ownership, promotes social cohesion, and

R. Gulzar · S. Gulzar (✉) · S. Mujahid
Department of Architecture, University of Management and Technology, Lahore, Pakistan
e-mail: dr.saimagulzar@yahoo.com

R. Gulzar
e-mail: rafia.gulzar@arch.uol.pk

R. Gulzar
School of Architecture, University of Lahore, Lahore, Pakistan

A. Cheshmehzangi et al. (eds.), *Nature in Cities, Nurturing Cities*, Urban Sustainability,
https://doi.org/10.1007/978-981-95-5487-4_8

provides opportunities for local employment. In conclusion, this nature-driven placemaking strategy for Sheikhupura, through the revitalization of existing green areas and the development of future green spaces, offers a holistic approach to combating urban decay.

Keywords Climate resilience · Nature-driven · Pakistan · Placemaking · Urbanization

1 Introduction

The expansion of cities is a global phenomenon that occurs mainly due to economic reasons. The current exponential trend of cities' growth project 70% of the world population in cities by 2050 (Adger et al. 2020; Cichowicz and Bochenek 2024; Pison 2022). This global increase in the urban population has resulted in unplanned and unregulated urbanization which is disrupting the economic and ecological ecosystems of the cities, while driving towards urban decay (Adger et al. 2020; Cichowicz and Bochenek 2024; Pison 2022). The decay emerges through a combination of pressures including overstretched infrastructure, inadequate services, unregulated land use, and the depletion of natural resources, resulting into poor living conditions. As cities are struggling to accommodate this exponentially growing population, overutilisation of basic services such as water, energy, waste management, and public health infrastructure leads to deteriorating living conditions. This rapid rate of urban decay around the globe is leading towards unsustainable areas with insufficient facilities as well as social and environmental inequalities in addition to loss of natural green areas for urban resilience (Adger et al. 2020; Cichowicz and Bochenek 2024; Pison 2022).

The unplanned growth and economic instability are the major driving factors for urban decay in the cities of developing countries, such as Pakistan (Khan et al. 2016; Ministry of Climate Change 2022; Patel and Suthar 2022). The rapid increase in the urban population due to rural–urban migration in Asian context has resulted into the haphazard growth of cities beyond the urban fringes in Pakistan (Khan et al. 2016; Ministry of Climate Change, 2022). This unplanned expansion usually lacks the basic urban infrastructure including clean air, water and energy resources, housing, as well as public health and transportation, leading towards urban inequalities. These challenges require mitigation through nature-based solutions for the rapidly urbanizing landscape of the city Sheikhupura in Pakistan to overcome the climate vulnerabilities, resource depletion, and habitat loss (Malik et al. 2012). Nature-based solutions are based on sustainable approaches that not only address the environmental and infrastructural deficiencies caused by urban decay but also nurture social well-being by producing healthier, more inclusive, urban environments.

Urban decay is the result of a complex interplay between demographic shift, inadequate planning, and ecological degradation that contributes to physical deterioration. Furthermore, it causes the breakdown of the sociocultural fabric of communities. In this scenario, transformative placemaking surfaces as a strategic response to urban decay, since it can transform the decaying spaces through the principles of community engagement, inclusive design, and sustainable practices.

Transformative placemaking is a multidimensional concept used for the creation of spaces that are not only sustainable but at the same time respond to community needs by facilitating their connectivity and restoring their sense of belonging (Adger et al. 2020; Cichowicz and Bochenek 2024; Pison 2022). The inclusivity and sustainability are the main elements in addition to the economic viability. The traditional placemaking techniques implemented in the past were mainly focused on the superficial aesthetic treatments without context and economic factors (Adger et al. 2020; Cichowicz and Bochenek 2024; Pison 2022). However, the transformative placemaking is aimed at the engagement of communities in reshaping their environments. It is rooted in social justice, equity, and inclusivity to create resilient spaces.

The traditional placemaking approach involves communities in shaping the space with their ownership while restoring cultural identity and the local context through participatory urban design as a main tool. The process of participatory urban design is usually conducted through connecting with the local residents and chalking down their requirements (Adger et al. 2020; Cichowicz and Bochenek 2024; Pison 2022). The community participation ensures long-term care and stewardship in addition to the sense of ownership. This aims to create spaces that are not only aesthetically appealing but also embedded with social and cultural context. Transformative placemaking is all about nurturing the holistic sense of place while incorporating elements, such as affordable housing, urban green spaces, as well as public art and accessible transportation (Adger et al. 2020; Cichowicz and Bochenek 2024; Pison 2022). Furthermore, this is one of the most powerful tools to address the issues of urban decay and environmental degradation while ensuring the community involvement in decision-making for sustainable urban growth.

Transformative placemaking is a dynamic approach. This approach is used to create environments that nurture well-being, connectivity, and sustainable growth. The well-being (physical, mental, and social) is embedded in nature and its interconnectivity also shapes the economic reasoning for the nature-driven placemaking (Project for Public Spaces 2020). The concept is not only about the reshaping of physical attributes of the space, rather fostering the social fabric of the community with NbS for well-being (Courage et al. 2021; Evans 2002; Friedmann 2007).

Sheikhupura, Pakistan, like many other urbanizing cities, is encountering various problems related to urban decay, environmental degradation, and the erosion of habitats. Nature-driven placemaking, involving the integration of ecological, social, and economic factors, is the only solution to create sustainable, resilient, and inclusive urban landscape in the city of Sheikhupura.

2 Transformative Nature-Driven Placemaking

Transformative placemaking is characterized as a people-centric approach. Placemaking in urban design is a tool that creates engaging and vibrant environments while establishing the deeper connectivity among all the community members for sustainable growth. The existing scenario of rapidly evolving urban landscape is converting the greens into greys by accommodating the exponential growth (Courage et al. 2021; Friedmann 2007). Therefore, the need for liveable spaces that provide social interaction, encourage community bonding, reflect cultural identity, as well as ensure safety in addition to human well-being, is increased. This tool transforms the places into liveable ones by enhancing their connectivity patterns, designing aesthetically appealing vibrant activities, and accommodating all sectors of society with diverse backgrounds. This bridges the trust through walkability and permeability while attracting the economic benefits for sustainability (Project for Public Spaces 2020). The aim is to make cities as the hub of activities to live, work, and enjoy.

The International Union for Conservation of Nature (IUCN) defines Nature-based Solutions (NbS) as strategies focused on the protection, sustainable management, and restoration of natural as well as modified ecosystems (Bush and Doyon 2019; Cannas et al. 2018). These mitigation measures are chalked down for nurturing biodiversity in addition to the enhancement of human well—being. Furthermore, NbS provides strategic measures for the adverse climatic and environmental threats (Faivre et al. 2017).

Although distinct, both transformative placemaking and nature-based solutions (NbS) aims to create resilient, inclusive, and livable environments. Transformative placemaking empowers communities and nurtures a sense of place and belonging, while NbS enhances ecological health and adaptive capacity, grounded in local contexts and community stewardship. Their integration promotes not only spatial and environmental justice, but also social equity and inclusivity. The transformative placemaking, embedded with nature, has emerged as a fundamental strategy for the enhancement of urban spaces. The inclusion of natural elements, such as green spaces, water features, and biodiversity significantly improves the ecological sustainability and quality of life for inhabitants (Bush and Doyon 2019; Cannas et al. 2018). The disruption of natural environments during urbanization and infrastructure development requires a reconnection with nature to foster mental, physical, and social well-being of the inhabitants. NbS in placemaking refer to a multidimensional approach to address challenges related to environmental degradation. For instance, the urban green space not only improves the air quality and enhances the biodiversity, however, also provides space for recreation, facilitating the physical activities as well as social interactions among all sectors of communities (Fig. 1).

The growth/expansion of cities is an inevitable process that continues and requires transformative nature-driven placemaking (Piracha and Chaudhary 2022; Sinnett 2020; Weber and Sciubba 2019). This approach devises environments that are resilient, adaptable, and conducive to the well-being of all residents in addition to holistic solutions to the modern urbanization challenges. Unlike traditional

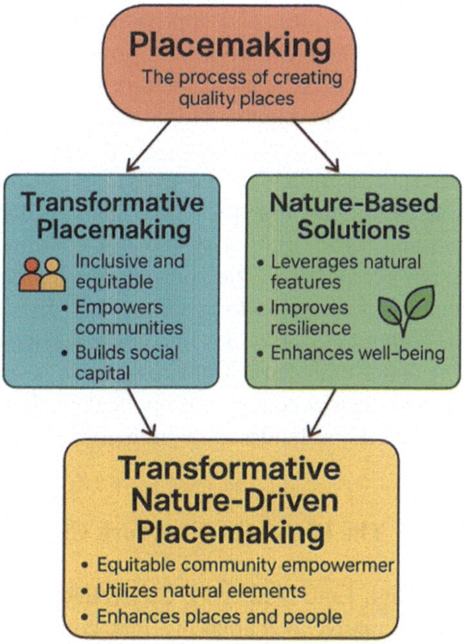

Fig. 1 Transformative nature-driven placemaking. *Source* The Autho)

approaches, this dynamic concept reimagines urban spaces as multifunctional hubs of activities including work, commerce, and recreation, while embedding economic prosperity within a framework of ecological sustainability (Bush and Doyon 2019; Cannas et al. 2018; De Ridder et al. 2008). Therefore, the scale for this well-defined scope is not limited to a public space rather requires larger areas with place-based challenges. This is an integrated approach that connects the diversified sectors such as economic development, community development, transportation, healthcare and more, in order to nurture local growth and progression through cohesive, nature-based weaved strategies.

2.1 Selected Case Studies for Transformative Nature-Driven Placemaking

Transformative nature-driven placemaking transforms the areas into more functional, safer, and accessible spaces by weaving connections between community and space. The whole concept is based on the fundamental principles of social inclusion, cultural identity, economic viability, and finally culminating into sustainable environments (Elmqvist et al. 2015; Pineda-Pinto et al. 2021). It gained attention in the last decade to accommodate the modern urbanization pressures and to uplift the quality of life in environmentally degraded areas in different parts of the world. Diversified case studies (Table 1) were selected for the comprehensive understating of different typologies, scales, and approaches to be interpreted for the city of Sheikhupura, Pakistan. The New York City project presented the case of abandoned and under-utilized blocked spaces, transformed into vibrant green urban parks/spaces with accessibility and inclusivity in addition to the preservation of heritage.

While studying the European perspective, the London future city project emphasized the creation of mixed-used neighbourhoods with green urban spaces. The Leeza Soho, Beijing, China presented business district urban square project rooted into the national transportation system for community-centric urban design to exhibit the Asian perspective. The rooftop development project in Dubai Mall, Dubai is the remarkable creation to accommodate diverse needs of community by transforming the commercial activities into engaging and resilient urban amenities with aesthetics. The Vanasthali Park, Lucknow, India is one of the exemplary projects that transformed the locked land within the densely populated neighbourhood into ecological and functional landscape with associative values evoking history and culture.

2.1.1 The High Line New York City

The High Line New York City reflects the breathtaking outcome of nature-driven transformative placemaking that converted an obsolete abandoned infrastructure into a vibrant public space. The project employed biophilic design while preserving the industrial heritage consisting of steel and concrete. The large-scale project was implemented in phases with public and private partnership. This case highlighted the abandoned railway track infrastructure and grey spaces being transformed into a joyful, socially inclusive, and sustainable environment with the implementation of NbS. The successfully implemented strategies, such as green and blue walkways, narrow and wide paths connecting parks, decked and open-air routes, high-end art district concept, green native plantings, various seatings with pedestrian viewing platforms, and green parking areas provided nature-based techniques for the chaotic unplanned growth of Sheikhupura with innumerable grey areas. The added economic benefits were observed in elevated prices of the neighbourhood that further motivated the surrounding community to participate in the project. The community enhanced their participation by extending the backyards of their property into poetic grunges garages.

2.1.2 The London Future City Project, UK

UK: The London Future City Project highlighted various options of different scaled projects (starting from a small urban square project of Gilt of Cain, London, UK to large-scale Battersea power station development project) that were transformed into vibrant public spaces with economic benefits in addition to promoting the eco-friendly environment while restoring their ecosystems. Various projects adopted the blue-green strategies comprising green infrastructure with water bodies for conversion of dull areas into liveable spaces by connecting the disrupted ecosystems and

Table 1 Case studies interpretation of transformative nature-driven placemaking in different regions

Case studies	Transformative placemaking principles	Strategies/Indicators	Nature-driven implementation
High Line in New York City, USA Challenges 1. Threats of demolition 2. Complex funding mechanisms 3. Engineering difficulties 4. Gentrification	Social inclusion	Accessibility	Elevated Promenade in the centre of thick tree plantations give an effect of Forest Reclining benches for Sun Bathing Pedestrian ways Green Elevators Green Ramps Open seating spaces, Creating viewing vistas, green oasis Steel and concrete preserved with greenery
		Inclusivity	
		Diversified seating	
		Interactive spaces for all sectors of society	
		Viewing galleries	
	Environmental sustainability	Greening of the area	
		Natural ventilation	
		Rainwater harvesting	
		Biodiversity	
	Cultural identity	Industrial Heritage	
	Economic Viability	Uplifting of neighborhoods	
		Provision of basic services	
		Gender specific design	
		Multiuse spaces	
The London future city project, UK	Social inclusion	Through Art displays	Vibrant public squares Seating for All Viewing spots User diversity Green infrastructure with blue water areas Green open public spaces, gyms, viewing platforms etc. in the vicinity of the power station Restoration of power station areas by adaptive reuse to further enhance the heritage value Space for informal economies Small entrepreneurs Vibrant mixed-use areas with residential apartments, offices spaces, retail shops
		Exhibition for All	
		Accessibility and Inclusivity	
		Design promoting community gatherings	
	Environmental sustainability	Pedestrian pathways	
		Electric vehicular Access to the Tourist areas	
	Cultural identity	Battersea Power Station obsolete structure restoration	
	Economic viability	Tourism	
		Commercial and Residential developments	
		Leisure spaces	

(continued)

Table 1 (continued)

Case studies	Transformative placemaking principles	Strategies/Indicators	Nature-driven implementation
Leeza Soho, Beijing, China Challenges 1. diagonally bisecting subway tunnel, which necessitated complex structural engineering	Social inclusion	Community centric approach	Bringing the community in the central atrium directly connected to the city's transportation network Splitting the tower into two halves around the tunnel, with a central atrium acting as a public square directly connected to the city's transportation network Two towers reflection of balance of opposites Traditional Architecture
		Escalators	
		Atrium (accessibility/inclusivity)	
	Environmental sustainability	Vertical park featuring waterfalls, walking paths, effectively bringing nature into the heart of the building	
		LEED certification	
		Natural and daylight usage	
	Cultural identity	The central atrium, reminiscent of a traditional Chinese courtyard, serves as a communal space, fostering interaction and connection among occupants	
	Economic viability	Increased influx of business due to accessibility and inclusivity	

(continued)

Table 1 (continued)

Case studies	Transformative placemaking principles	Strategies/Indicators	Nature-driven implementation
The Dubai Mall, UAE Challenges: 1. Retractable roof capable of opening during pleasant weather while withstanding high winds, heavy rain, and sandstorms typical of Dubai's climate 2. Cost–Benefit Analysis 3. Energy efficiency	Social inclusion	Accessibility	VR park, children's play area, and an Olympic-sized skydiving chamber Greening of retail shops Rooftop as park
		Inclusivity	
		Diversified seating	
		Interactive spaces for all sectors of society	
	Environmental sustainability	Energy Efficient design	
		Biodiversity	
	Cultural identity	Traditional shops and cuisine	
	Economic Viability	Commercial and retails services	
Vanasthali Park, Lucknow, Uttar Pradesh, India Challenges: 1. Fragmented land 2. locked with densely plotted development 3. Garbage disposal space 4. Waterlogged areas	Social inclusion	Accessibility	Public Park serving to all surrounding residential Vanasthali forms an ecological and functional landscape With associative values evoking history and culture in a stylised contemporary language Abandoned place into tourist attraction
		Inclusivity	
		Various groups in the neighbourhoods	
	Environmental sustainability	Restoration of ecosystem	
	Cultural identity	Rehabilitation of Fortress	
	Economic viability	Tourism base economy	

providing accessibility in addition to adaptive reuse strategy. The urban square project of Gilt of Cain, London displayed a unique combination of art, poetry, architecture, and public space. The Battersea power station project was developed in phases starting from the vibrant mixed-use areas with residential apartments, office spaces, retail shops, green open public spaces, gyms, and viewing platforms in the vicinity of power station. The second phase was the redevelopment of the iconic buildings of power station into boutiques, hotels, retail areas, and cafes in addition to penthouse with garden square above the boiler house. Sheikhupura is an industrial city and its expansion without a master plan in different times created dark and dull spaces. Such unplanned and unchecked developments are currently acting as a central core to criminal activities. This is one of the reasons that Sheikhupura is considered unsafe and dangerous while travelling most of the times. The creation of safe and liveable spaces in London Future City Project provides exemplary options for interpretation in the context of Sheikhupura city.

2.1.3 China

The Leeza Soho, Beijing, China is the finest interpretation of fluidity concept of Zaha Hadid architects. The beautifully carved transformation of a major Leeza Soho office hub with a transport substation into an urban square is directly connected to a transport system in the form of Atrium with vertical gardens and waterfalls. This community-centric space enhances the overall experience through green walks surrounded by waterfalls and promotes accessibility to all. The leftover spaces within the buildings due to linear urbanization is one of the major problems in the cities of Pakistan due to limited resources. The haphazard urban growth in Sheikhupura has led to the development of such spaces in abundance which needs a thorough fluid design to connect such spaces. This implementation of connectivity with natural fluidity would enhance the quality of life in Sheikhupura.

2.1.4 Dubai

The Dubai Mall, UAE Rooftop Development Project transformed the commercial hub into the family entertainment destination in order to fulfil the requirements of diverse communities. The purpose was to create an urban resilient amenity through technological innovation embedded with NbS to mitigate the climatic extremes of Dubai. The technology played a vital role by accommodating the retractable roof offering various options according to the climate in addition to the energy efficiency. This high-tech solution would be a resource incentive in the context of Sheikhupura, however, rooftop developments with NbS might regenerate the degraded ecosystems and enhance the biodiversity. The large industrial units bared roofs can be utilized for green foliage to improve the overall environmental quality and may act as a purifying agent.

2.1.5 India

The Indian Project, that is, Transforming Urban Landscape, demonstrated various projects including river fronts, urban spaces, transit nodes, streetscapes, conservation adaptive reuse, and parks through placemaking designs in different parts of India. The Vanasthali Park in Lucknow-India was a deserted triangular piece of land being transformed into teak plantation trail along the winding dry-stream. The low-lying water-logged area was channelized as lake and the distorted perimeter due to dense population in the surroundings was restructured with historical significance of fortress. The neglected and abandoned locked land due to densely populated surroundings along with the fortress was revived as a forest heaven. Sheikhupura is a historical city with numerous heritage sites that require a serious attention due to the densely populated surrounding areas. The expansion with lack of infrastructure is eroding the heritage sites and converting them into crumbling structures as well as forwarding them towards demolition.

2.2 *Potential of Nature-Based Placemaking in Addressing the Challenges Faced by Sheikhupura City*

The Sheikhupura city (Fig. 2) lies at a latitude of 31° 42′ 36″ North and a longitude of 74° 08′ 00″ South. The city is well connected to major surrounding urban centres of Lahore, Faisalabad, Sargodha, and Gujranwala. The city was planned with zones by placing the grain and vegetable markets along with the retail spaces in the centre. The government and public institutions were planned within the grid-iron pattern towards west comprising municipal offices, courts, heathcare buildings, and gardens (Ashraf et al. 2024). The demographic shift after 1947 expanded the city haphazardly in the southern and eastern directions. The unplanned expansion with an increase in the population changed the overall shape of the planned city. The uncontrolled growth patterns disrupted the natural ecosystem as well as anthropogenic infrastructure. This forced urbanization with the influx of population and converted most of the open areas within the city into densely populated and narrowed streets and haphazard residential buildings without the basic infrastructure facilities (road network, water and gas supply, sewerage system, and garbage disposal).

One of the most pressing issues arising from unplanned urban growth is an increase in vehicular traffic, which, coupled with the growing population density, has led to severe congestion and air pollution (Ahern et al. 2014; Andersson et al. 2014; Imam and Banerjee 2016). Overcrowded residential areas, often characterized by informal settlements, have also emerged as a consequence of this haphazard growth. These settlements, which are typically developed without adequate infrastructure or urban design, are characterized by poor living conditions and a lack of basic services, such as water, sanitation, and proper waste management. Such hazards initiate the pattern of

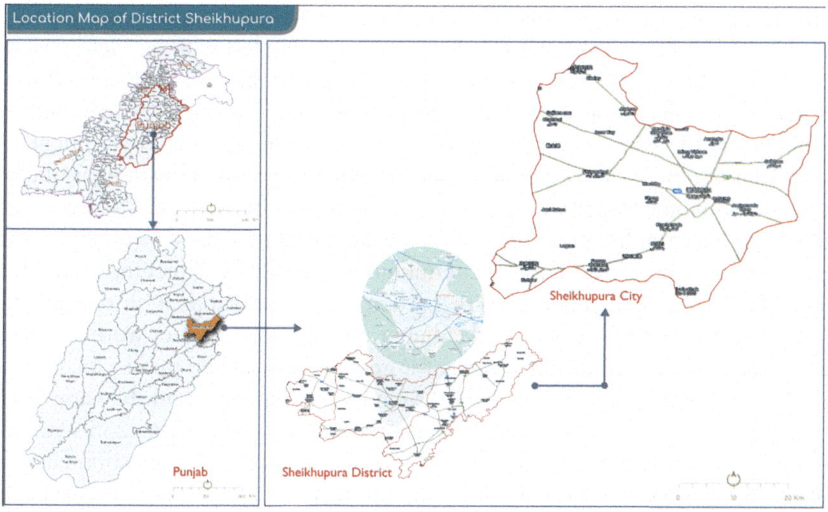

Fig. 2 Location map of study area Sheikhupura city-Pakistan. *Source* The Author

decline that culminates into urban decay in the broader perspective. Unregulated and unplanned urban growth usually remains unaddressed, leading to the deterioration of urban systems, environmental degradation, and escalation in social inequality. All these are the main contributors, rather accelerators in the process of urban decay.

Moreover, encroachments on public land and the absence of effective regulatory enforcement have aggravated the situation, contributing to the city's deteriorating condition (Ahern et al. 2014; Andersson et al. 2014; Imam and Banerjee 2016). There is an urgent need for detailed urban design/planning strategy based on sustainable growth and development to mitigate the significant challenges for resilient and inclusive future.

3 Methodology

The current research aimed to explore the potential of Sheikhupura city for the implementation of nature-based placemaking in the urban context by employing the mixed-methods approach (Fig. 3). Comprehensive studies were conducted through both qualitative and quantitative techniques to develop the baseline data for the application of NbS. The sequential process initiated with the literature review for the comprehensive review of case studies followed by field data acquisition through questionnaires and visual surveys at different locations in Sheikhupura city. The detailed observations combined with field data were then compiled for the selection of appropriate NbS for the targeted areas.

The literature review was conducted to understand the existing urban decay scenario of Sheikhupura city by navigating through similar cities around the globe. The comprehensive analysis listed the challenges and opportunities in addition to the varied appropriate interpretations of NbS utilized in transformative placemaking in different parts of the world. This was further complemented with the field observational data for the identification of sites. The qualitative data was collected through interviews and focus group discussions. Semi-structured interviews were conducted with all stakeholders including urbanists, environmentalists, local government officials, and community to record their perspectives for the existing challenges of urban decay, environmental degradation, and loss of green public spaces. They were also asked for the NbS implementation in the existing scenario. Focus group discussions were organized with the residents of communities so that their direct input should be incorporated in the final proposal. The participatory workshops with participatory mapping, brainstorming sessions, and collaborative design exercises were concluded with the integration of community ideas for green corridors and public spaces.

The quantitative data was collected through questionnaires and the sample size was derived from Slovin's formula. The sample size calculated was 235 with 95% of confidence level to achieve the objectives of the research. The questionnaire was divided in three parts to collect all the required information including their socioeconomic status, perspectives of urban decay, and the requirements for green infrastructure. The set of requirements derived was then designed as a model for interventions

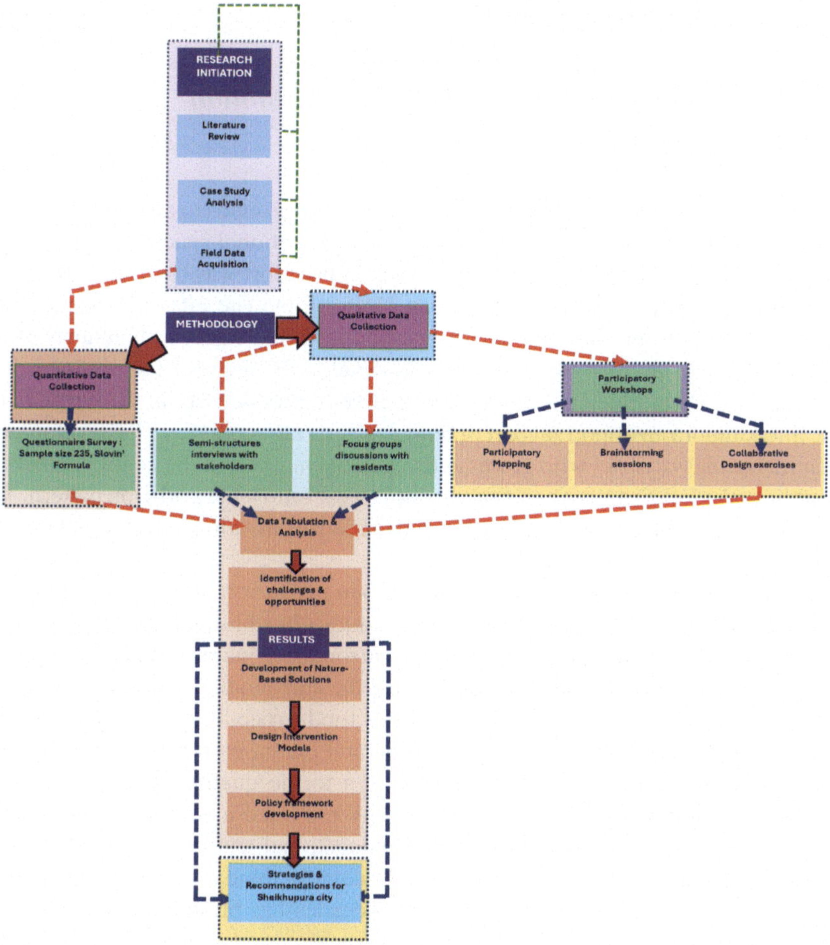

Fig. 3 Research methodology flow chart. *Source* The Author

that may be replicated in different parts of the city and finally the policy framework to be generated accordingly. The study concluded by highlighting strategies and policy recommendations for Sheikhupura city.

4 Results and Discussions

4.1 Historical Context and Urban Transformation of Sheikhupura

The city of Sheikhupura evolved from Mughal hunting grounds to an agricultural hub and finally to industrial city. It was founded as a recreational space with lush green areas in addition to hunting grounds. The Sheikhupura fort was built in 1607 during the reign of Emperor Jahangir (The urban Unit 2021). The city transformed into agricultural hub during the British colonial period in the eighteenth century. They constructed canals and irrigation systems to develop the area for the production of crops, such as wheat, rice, and cotton. The increased productivity in the fertile lands made Sheikhupura an important part of Punjab's agrarian economy. The development of central business district with grains and vegetable markets in the centre along with the public and administrative buildings towards west follow the grid-iron pattern. The rapid growth after independence in 1947 due to industrialization transformed the city into concrete jungle.

The Sheikhupura city has seen informal urban expansion after 1947 mainly due to the industrial growth. The economic opportunities in addition to the well-planned infrastructure and interconnectivity with the surrounding major urban areas uplifted the city as the most favourable destination for the surrounding areas. Considerable investments in the industries, such as textile, food processing, and pharmaceuticals after 1960 caused an influx of population and increased rural–urban migration phenomenon for better employment in addition to the quality of life (The urban Unit 2021). The basic infrastructure of the city provided a well-connected public transport system, road networks, as well as educational and medical facilities that served as the basis for the above-mentioned demographic change in that period. Further haphazard expansion in the last two decades has brought challenges, such as rapid population growth, inadequate infrastructure, and socioeconomic disparities in addition to informal settlements, environmental degradation, and governance challenges.

Sheikhupura city is deteriorating at a faster pace since the last two decades as the demands for the increased population remained unfulfilled. The over consumption of city resources in addition to the unplanned expansion converted most of the developed land into debris. The increased vehicular traffic, over-populated areas, and encroachments have taken away most of the green areas of the city. The informal settlements acquired the land without infrastructure facilities that with the passage of time converted into slums and ruined the overall built environment of the area. The selected area comprised of CBD, Civil lines, Old City, Fort and Agriculture Farmlands as shown in (Fig. 4) was studied in detail to implement the transformative nature-driven placemaking.

Sheikhupura is currently facing significant challenges related to unplanned growth and urban decay due to rapid urbanization. The uncontrolled densification (encroached informal settlements on green open spaces) of already populated areas,

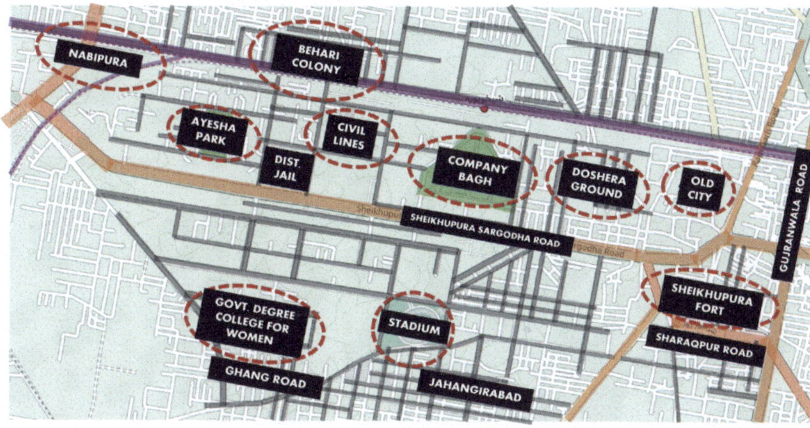

Fig. 4 Selected segment for application of transformative nature-driven placemaking. *Source* Author site identification on google maps

such as Bihari Colony, Nabi Pura, and Doshera ground) without basic urban infrastructure in the past two decades is one of the major contributors to the depletion of the resources and creation of inhabitable spaces. These encroachments along major road arteries increased the vehicular traffic with increased air and noise pollution levels in the central business district in addition to congestion and inaccessibility. The overcrowding and congestion are declining the local businesses in the vegetable and grain market along with the main bazar/markets. Furthermore, the economic instability is generating unemployment and discomfort for the community. This mismatch between population growth and resource availability has contributed to the deterioration of the urban environment, as the city's infrastructure failed to accommodate the increasing demand. Urban parks, such as Ayesha Park, Company Bagh, and Liaquat park are shrinking and informal residential/commercial activities are increasing. This unchecked expansion of city has resulted in a significant loss of green spaces and encroachment of natural habitats while diminishing the city's aesthetic appeal and environmental sustainability.

The observational survey of the identified areas in the selected segment as shown above highlighted the impacts of urban decay (Table 2a) majorly in loss of biodiversity and green spaces. This situation was further complemented by health and safety issues in these areas. The deterioration of environmental quality enhanced with the disruption of ecological systems that converted once green lands into heaps of debris. The broken ecosystems closed the ecological corridors and destroyed the natural habitats resulting into decline of native plants and animals.

Sheikhupura city, known for its lush green lands with spouting waters, native plantations, singing birds, and diversified wildlife, has been totally transformed into a dull and underutilized/abandoned area full of debris, core of health problems, and over-crowded without safety. All these factors contributed in the lack of community interest and ownership in addition to lack of control by the government. The

Table 2 Assessment for transformative nature-driven placemaking through (a) observational and (b) stakeholder questionnaire survey in Sheikhupura

(a) Assessment (through observational survey in selected areas)	Typologies	Parks	Main bazar	Residential	Informal settlements	Old city and fort
Attributes	Natural feature	Yes	No	Yes	No	Yes
	Usage patterns	Not defined	Defined	Defined	Not defined	Not defined
	Accessibility/inclusivity	No	No	No	No	No
	Health and safety	No	No	Yes	No	No
	Hard and soft landscape	Yes	Yes	Yes	No	To some extent
	Cultural and historical integration	No	No	Yes	No	Yes
	Evidence of community stewardship	No	No	No	No	No
(b) Assessment (through stakeholder questionnaire survey)	Concept of nature-driven placemaking	Perception	Satisfaction	Awareness	Suggestions	Engagement
Stakeholders	Local residents	No	No	To some extent	Ready for participation	No
	Business owners	No	No	No	Not sure for participation	No
	Government officials/ Professionals	Yes	No	Yes	Implementation agency	No
	Environmental/NGOs/ Heritage	Yes	Yes	Yes	Ready for participation	Yes
	Youth groups	To some extent	To some extent	To some extent	Ready for participation	Yes

questionnaire survey (Table 2b) from different stakeholders including local residents, business owners, government officials/professionals, NGO's, and the youth about nature integration as the solution for controlling the deteriorated built environment emphasized collaborative efforts. There is a need for awareness first as all sectors of society are ready to participate and contribute for the upgradation of living quality/standards. Moreover, there is s need of a framework through which all sectors can play their role and their willingness for contribution further strengthened the preparation of design proposal for Blue-green Trail (BGT) to Sheikhupura Fort. This design also required assessment (physical, spatial, environmental, and socioeconomic attributes) of the existing characteristics of green areas and infrastructure evaluation for implementation as shown in Table 3.

4.2 *Blue-Green Trail (BGT) to Sheikhupura Fort*

The assessment of urban landscape revealed a critical deficiency in availability, accessibility, and quality of basic features and green areas (Andersson et al. 2016; Ziaf and Hyder 2022). The parks and open spaces are few in number with poor maintenance and uneven distribution. The hard and soft landscape features are mostly in deteriorated condition and heaps of debris covered most of the neglected areas. The roads and streets lack greenery and pedestrian comfort and are one of the main contributors in environmental pollutants. The public squares and markets are congested, mostly encroached, polluted, and devoid of green integration (Ahmad and Anjum 2020; Ahmed et al. 2020). These encroachments stretched all over the open and green spaces within the commercial areas and contributed to environmental stress to its inhabitants. The old planned residential areas are in better conditions but the urban sprawl in the form of informal settlements encroached all buffer areas and are facing severe green deprivation. The heritage sites are under severe environmental stress due to neglect, encroachments, and lack of green buffer zones. These results further complemented the observational and questionnaire survey by verifying the contribution of all these stated factors in environmental degradation, reduced quality of life, increasing heat impacts, poor air quality, and loss of cultural identity and liveability in Sheikhupura city (Fig. 5a). Blue-Green Trail (BGT) initiative has been conceptualized to address these issues and implemented as a nature-based solution (NbS) for weaving the neglected and fragmented urban spaces into lively ecological corridors, embedded with cultural and economic perspectives (Fig. 5b).

The Blue-green trail (BGT- an urban ecological spine) to Sheikhupura Fort starts from Gateway in Nabipura where the fragmented, underutilized, and landlocked spaces were utilized to formulate a welcoming starting point for this trail. These spaces in the Nabipura were abandoned due to closer vicinity to railway track and were declared unsafe for travel route (waste heaps and debris). However, this was main road junction in the past with greenery, in particular flora and fauna species. The beauty of this trail is threading of deserted green areas that once stood isolated, fragmented, and in some cases encroached along the route. This transformation is based

Table 3 Availability assessment of basic features in selected areas and their characteristics for Blue-Green Trail to Sheikhupura Fort

Place	Attributes	Yes/No
Parks	Accessibility/inclusivity	No
	Functionality	Yes
	Number and area satisfactory	No
	Condition satisfaction (plantations, ecological features etc.)	No
	Usage	Yes
	Safety	No
	Water bodies	No
Roads/streets	Roads and street greenery	No
	Infrastructure design and maintenance	No
	Connectivity to green areas	No
	Encroachments	Yes
	Drainage and waste management systems	No
Markets	Integration of greenery	No
	Basic infrastructure aligned with greens	No
	Ventilation and shade (air circulations, buffer zones etc.)	No
	Environmental load (pollution, noise etc.)	Yes
	Spatial efficiency (open vs covered areas, parking etc.)	No
Squares	Presence of natural features	No
	Functionality (mixed use commercial, cultural and social)	No
	Physical condition acceptable	No
	Accessibility/Inclusivity	No
	Cultural values	No
Residential	Green space access	Yes
	Garden	Yes
	Street trees and community gardens	No
	Environmental quality satisfactory	No
	Urban Sprawl impacts	Yes
Informal settlements	Access to green areas	No
	Land use pressure	Yes
	Environmental hazards	Yes
	Open space usage	No
	Liveability indicators	No
Heritage sites	Landscape integration	No
	Preservation of natural surroundings	No
	Encroachments and Pollution stress	Yes
	Tourism	No
	Buffer zone with greens	No

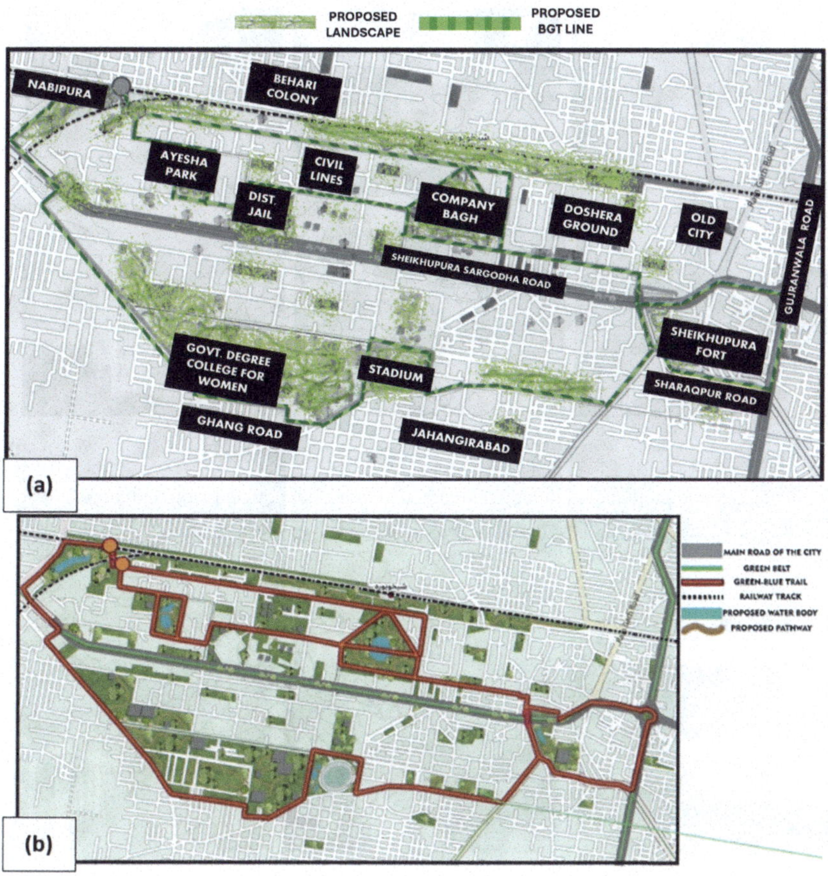

Fig. 5 **a** Conceptualization of Blue-Green Trail to Sheikhupura Fort. **b** Connecting the ecological features with Urbanization parameters to formulate the sustainable, resilient and inclusive ecological urban landscape. *Source* Author

on the principles of Nature-based Solutions which aim to restore the degraded built environment while providing socioecological benefits (Badach et al. 2022; Mahmoud and Morello 2021). The green connecting pathway is a threading of deserted areas into urban healthier living ecological corridors from Ayesha Park (Fig. 6a and b) to Company Bagh (Fig. 6c and d).

The existing park deteriorated condition reflects loss of green areas and their rapid expansion into the barren patches (Fig. 7a) leading to accelerated loss of biodiversity (both flora and fauna). This neglect is further complemented with the unchecked public usage creating unhealthier and shady spaces with security concerns. Therefore, the proposed design (Fig. 7b) of organically planned green areas, water bodies, and green permeable paved pathways as theme of BGT trail route is the sustainable solution to mitigate the deteriorated built environment of the existing parks.

Fig. 6 **a** Ayesha Park existing scenario, **b** proposed implementation of native plantations, permeable pathways and water to regenerate Ecological Urban Landscape, **c** Company Bagh existing condition, and **d** proposed addition of green connecting pathways. *Source* Author

Similarly, there are many other deserted areas along the connecting pathways that requires immediate transformation. These areas were mostly green and open spaces that transformed into heaps of debris with time and further deteriorated the built environment. Therefore, the whole BGT route is injected with native plantation, permeable pathways, and water punctures to improve stormwater management, enhance biodiversity, lower urban heat, and control air pollution in order to create liveable spaces for the community. These include viewing vistas, spaces for art and cultural displays, play areas, jogging tracks, seating islands, blue oasis, and street cafes, along with the restoration of arts, crafts, and flora and fauna. These not only enhance the recreational and cultural value of such spaces but also contribute to mental health and social cohesion.

The spaces are planned for all age groups, such as seating corners for women gossiping, exercise areas for healthy lifestyle, nooky hideouts for children to play,

Fig. 7 **a** Existing connecting pathway with deserted areas, **b** proposed nature embedded recreational spaces along the proposed winding BGT route to nurture the community and space for well-being. *Source* Author

and grassy alcoves for older groups. These enjoyable spaces for all sectors of society embedded into historical and cultural context are accessible and inclusive which nurtures the relationship with nature (Kalfas et al. 2023; Kang et al. 2024). This winding BGT route is nurtured with nature to halfway destination of Sheikhupura Fort heritage site. The proposed green buffer zone around the Fort (crumbling due to environmental stress) acting as a protective ecological belt would preserve the heritage site and reduce further impacts by air/noise pollution, extreme heat and overcrowding (Coombes and Viles 2021). The historic and cultural surroundings of the Fort are also conserved with greener places to attract more tourists in addition to public facilities.

The economic degradation is controlled through provision of food stalls, arts and crafts shops, local entrepreneurships, and branding of traditional goods. Throughout the BGT route, the repairing strategies for ecological corridors of disrupted ecosystems through native plantations, breeding spaces for local flora and fauna are drafted. The returning route passing through cricket stadium grounds is also planned with greener community spaces. The residential deserted open spaces along the BGT route are also embedded with green and blue elements to revive the flora and fauna. The reintroduction generates the habitat formations to promote urban resilience. One of the significant features on the way back is experiencing the agricultural farms (once Sheikhupura is known for). The incorporation of this activity enhanced the beauty of this trail. The agricultural touch in the end not only adds to the unforgettable memories rather enhances the economic viability at the same time. The Blue-Green Trail is a nature-based intervention for the selected urban segment embedded with ecological connectivity, climate adaptation, cultural identity, and an inclusive public space design.

5 Conclusions

The current study emphasized the critical role of the transformative, nature-driven placemaking to address the challenges of unplanned urbanization in Sheikhupura, Pakistan. The urban ecological network design focusing on the revitalization of the existing green area as suggested and the thoughtful planning of future public spaces, the city may move towards a more ecologically balanced, socially inclusive, and sustainable urban environment. NbS (permeable surfaces, native plants) with the integration of community-led planning and participatory process provided innovative solutions to counter urban decay with the enhancement of biodiversity for the restoration of public life in the city. The findings highlighted that the implementation of meaningful transformation is not only through infrastructure improvement but embedded in shift for governance and community engagement by fostering the local voices shaping the designs and stewardship of their urban ecological landscapes. The vision for greener Sheikhupura depends on long-term commitment, collaborative action, and the integration of nature as a central element of urban resilience and liveability.

6 Policy Framework

Vision

To transform Sheikhupura-Pakistan into a resilient, inclusive, and ecologically vibrant city by embedding nature-based solutions into urban design through community participatory process.

Objectives

- To restore ecological balance through revitalization of identified green spaces.
- To prevent urban decay through integration of nature-based infrastructure.
- To promote social inclusion by channelizing community participation.
- To enhance urban liveability through connected green networks.
- To foster economic resilience by green employments and eco-tourism.

Policy Actions

Action	Goal	Instruments for achieving the goal
Green space revitalization	Revitalization of existing green areas, enhancing parks, forests areas and public lands	• Identification for existing and future areas • Allocation of budgets for parks restorations and biodiversity enrichment • Implementation of native tree plantings, rewilding and soil restorations • Transformation of underutilized lands into accessible green commons
Inclusive urban design	Green spaces design that allows all sectors of society	• Integration of local communities through participatory design • Incorporation of universal design standards for accessibility • Prioritization of women, children, elderly, and marginalized groups of society • Ensuring spatial equity through distribution among all neighborhoods
Green connectivity and infrastructure	Interconnected urban nature network establishment	• Development of green corridors, walking tracks and Cycle ways • Integration of green roofs, rain gardens and bioswales in planning codes • Linking the public parks, schools and community areas through green pathways
Community stewardship and capacity building	Fostering local ownership and sustainable management	• Creating community-led green councils for areas • Training the youth as well as women for ecological maintenance • Incorporation of green education • Offering micro grants for community-led green projects

(continued)

(continued)

Action	Goal	Instruments for achieving the goal
Policy integration and governance	Embedded nature-based placemaking into citywide governance	• Integration of NbS into Sheikhupura's master plan and building bye-laws (environmental and social KPI's) • Developing public–private partnerships (PPPs) for green infrastructure

These policy actions require funding mechanisms (national and international or local community investments) in addition to monitoring and evaluation/feedbacks for successful implementations. Similarly, timeline for each action should be determined accordingly.

References

Adger WN, Crépin AS, Folke C, Ospina D, Chapin FS, Segerson K, Seto KC, Anderies JM, Barrett S, Bennett EM, Daily G, Wilen J (2020). Urbanization, migration, and adaptation to climate change. One Earth 3(4):396–399

Ahern J, Cilliers S, Niemelä J (2014) The concept of ecosystem services in adaptive urban planning and design: A framework for supporting innovation. Landsc Urban Plan 125:254–259

Ahmad N, Anjum G (2020) Green infrastructure: opportunities for urban resilience in Pakistan. J Sustain Dev 13(3):1–15

Ahmed A, Rana T, Khan AA (2020) Green growth strategy for sustainable urbanization in Pakistan: a system dynamics approach. J Environ Manag 268:110670

Andersson E, Barthel S, Borgström S, Colding J, Elmqvist T, Folke C, Gren Å (2014) Reconnecting cities to the biosphere: stewardship of green infrastructure and urban ecosystem services. Ambio 43:445–453

Andersson K, Dickin S, Rosemarin A (2016) Towards "sustainable" sanitation: challenges and opportunities in urban areas. Sustainability 8(12):1289

Ashraf T, Aslam J, Mehmood MS, Ahamad MI, Rehman A (2024) Geospatial assessment of built environment on land surface temperature in district Sheikhupura, Punjab Pakistan. Discov Geosci 2(1):30

Badach J, Szczepański J, Bonenberg W, Gębicki J, Nyka L (2022) Developing the urban Blue-Green Infrastructure as a tool for urban air quality management. Sustainability 14(15):9688

Bush J, Doyon A (2019) Building urban resilience with nature-based solutions: how can urban planning contribute? Cities 95:102483

Cannas I, Lai S, Leone F, Zoppi C (2018) Green infrastructure and ecological corridors: a regional study concerning Sardinia. Sustainability 10(4):1265

Cichowicz R, Bochenek AD (2024) Assessing the effects of urban heat islands and air pollution on human quality of life. Anthropocene 100433.

Coombes MA, Viles HA (2021) Integrating nature-based solutions and the conservation of urban built heritage: challenges, opportunities, and prospects. Urban For Urban Green 63:127192

Courage C, Borrup T, Jackson MR, Legge K, McKeown A, Platt L, Schupbach J (eds) (2021) The Routledge handbook of placemaking. Routledge, London

De Ridder K, Lefebre F, Adriaensen S, Arnold U, Beckroege W, Bronner C, Damsgaard O, Dostal I, Dufek J, Hirsch J, IntPanis L, Weber C (2008) Simulating the impact of urban sprawl on air quality and population exposure in the German Ruhr area. Part II: Development and evaluation of an urban growth scenario. Atmos Environ 42(30):7070–7077.

Elmqvist T, Setälä H, Handel SN, van der Ploeg S, Aronson J, Blignaut JN, Gómez-Baggethun E, Nowak DJ, Kronenberg J, de Groot R (2015) Benefits of restoring ecosystem services in urban areas. Curr Opin Environ Sustain 14:101–108

Evans P (ed) (2002) Livable cities?: Urban struggles for livelihood and sustainability. University of California Press.

Faivre N, Fritz M, Freitas T, De Boissezon B, Vandewoestijne S (2017) Nature-based solutions in the EU: innovating with nature to address social, economic and environmental challenges. Environ Res 159:509–518

Friedmann J (2007) Reflections on place and place-making in the cities of China. Int J Urban Reg Res 31(2):257–279

Imam AU, Banerjee UK (2016) Urbanization and greening of Indian cities: problems, practices, and policies. Ambio 45:442–457

Kalfas D, Kalogiannidis S, Chatzitheodoridis F, Toska E (2023) Urbanization and land use planning for achieving the sustainable development goals (SDGs): a case study of Greece. Urban Sci 7(2):43

Kang H, Fu M, Kang H, Li L, Dong X, Li S (2024) The impacts of urban population growth and shrinkage on the urban land use efficiency: a case study of the Northeastern region of China. Land 13(9):1532

Khan MA, Khan JA, Ali Z, Ahmad I, Ahmad MN (2016) The challenge of climate change and policy response in Pakistan. Environ Earth Sci 75:1–16

Mahmoud I, Morello E (2021) Co-creation pathway for urban nature-based solutions: testing a shared-governance approach in three cities and nine action labs. In Smart and sustainable planning for cities and regions: results of SSPCR 2019—open access contributions, vol 3. Springer International Publishing, pp 259–276

Malik SM, Awan H, Khan N (2012) Mapping vulnerability to climate change and its repercussions on human health in Pakistan. Glob Health 8:1–10

Ministry of Climate Change | Government of Pakistan (2022) *Pakistan's first biennial update report (BUR-1) to the united nations framework convention on climate change (UNFCCC)*

Patel A, Suthar A (2022) Analysis of urban development on land cover changes of three cities of Gujarat state, India. Geographica Pannonica 26(4).

Pineda-Pinto M, Frantzeskaki N, Nygaard CA (2021) The potential of nature-based solutions to deliver ecologically just cities: lessons for research and urban planning from a systematic literature review. *Ambio, 1–16*

Piracha A, Chaudhary MT (2022) Urban air pollution, urban heat island and human health: a review of the literature. Sustainability 2022(14):9234

Pison G (2022) World population: 8 billion today, how many tomorrows? Popul Soc 604(9):1–4

Project for Public Spaces (2020) What is placemaking?

Sinnett D (2020) Mitigating air pollution and the urban heat Island effect: the roles of urban trees. In *The Routledge handbook of urban ecology*. Routledge, pp 639–648.

The Urban Unit (2021) PCIIP cities profile: Sheikhupura. Government of Punjab

Weber H, Sciubba JD (2019) The effect of population growth on the environment: evidence from European regions. Eur J Popul 35:379–402

Ziaf H, Hyder L (2022) Expectation and reality of urban green space: a case study on landscape planning and development of Islamabad city, Pakistan. Discovery 58(321):941–952

Nature-Based Placemaking for Healing in Pacific Island Communities Post-disaster

Iderlina Mateo-Babiano, Judy Bush, and Xavier Cadorel

Abstract Natural disasters can cause a catastrophic impact on people and the places they love. A 2021 volcanic eruption and tsunami caused large-scale devastation affecting coastal communities including the Ohonua community in Eua, Tonga. Eua's foreshore became a no-build zone, and a place of trauma. The disaster not only eroded the community's sense of place, but also their social frameworks. This chapter presents an auto-ethnographic reflection of the potential role of nature in placemaking for healing. Through a Placemaking Sandbox travelling studio, twelve master-level students from an Australian university supported by four local students worked collectively with the Ohonua community to reimagine a happier and healthier way of living that can drive and support resilience and ecologies in Eua, while reducing community vulnerability, enhancing place attachment and belonging, and increasing disaster resilience to future shocks. In navigating the complex cultural settings of the Ohonua community, the students and community members co-designed nature-based opportunities and implemented four tactical placemaking initiatives that demonstrated biophilic, nature-based, and participatory design approaches to shape nature-centred public spaces, revealing important insights on the potential role of nature-driven placemaking in building back better. Moreover, it highlights how community groups can heal and thrive while increasing awareness of the economic, social, and cultural benefits of nature.

Keywords Nature · Place · Placemaking · Tonga · Island communities · Design for healing · Ecosystem services

---

I. Mateo-Babiano (✉) · J. Bush · X. Cadorel
Faculty of Architecture, Building and Planning, The University of Melbourne, Melbourne, Australia
e-mail: imateo@unimelb.edu.au

J. Bush
e-mail: judy.bush@unimelb.edu.au

X. Cadorel
e-mail: xavier.cadorel@unimelb.edu.au

A. Cheshmehzangi et al. (eds.), *Nature in Cities, Nurturing Cities*, Urban Sustainability,
https://doi.org/10.1007/978-981-95-5487-4_9

1 People and Place: The Pacific Island as Placemaking Context

The shaping of a climate-positive future for Indo-Pacific Island communities, particularly after a major disruption, demands full recognition that climate emergency and biodiversity loss are twin challenges that further compound the already catastrophic impacts of disasters on coastal people and the places that they love (UNEP 2021)

In 2021, the Hunga Tonga–Hunga Ha'apai volcano erupted in the Pacific Ocean, triggering a deadly tsunami. Coastal communities in small Pacific Island countries such as Tonga were seriously affected, including the Ohonua community in Eua Island. The tsunami caused large-scale devastation with many homes, buildings, roads, and ports damaged. It disproportionately affected the already vulnerable rural poor households in the Pacific region (Atkinson, 2023). In addition, livelihoods were lost. Eua's foreshore became a no-build zone, and a place of trauma for many families who also had to be relocated to another part of the island. Three years later, the emotional scars and place trauma remain. Many continue to express discontent and disconnection from their new homes. The disaster not only eroded their sense of place but also their social frameworks. While not a disaster of climate change, the earthquake and tsunami's ongoing impacts point to many lessons and insights relevant for both climate related and other disaster recovery processes.

There is already extensive documentation of the direct impacts of climate change, including coastal soil erosion, wetland loss, and escalating risks of natural hazards such as hurricanes/typhoons, storm surge and flash flooding, and the amplifying impacts of intersecting climate and non-climate related disasters. Not as obvious are the indirect effects of increasing levels of poverty and biodiversity loss. Acknowledging that these extreme events are becoming more frequent and severe, the need to address the intensifying vulnerabilities given a community's restricted access to innovations, infrastructure, and education becomes urgent and imperative (Dasgupta et al. 2014).

While many continue to struggle to discover pathways out of this emergency, education can take a leading role in inspiring learners and building community agency, referring to the act of coming together as a community to build a sense of place belonging, to shape more climate-resilient futures. Key to this is understanding our '*place*' as a space for healing.

In this chapter, we present an auto-ethnographic reflection of the authors/studio leaders supplemented with scholarly literature on the potential role of nature in designing for healing. Through the delivery of a placemaking travelling studio held at Eua Island, in the Kingdom of Tonga, students, in teams, worked in partnership with the Ohonua community in Eua to collectively reimagine happier and healthier ways of living that can drive and support resilience and ecologies. We contend that the place-based travelling studio can be understood as a 'nature-based Placemaking Sandbox model' (Mateo-Babiano and Palipone 2020a). The studio offers experiential learning and relational support in creating opportunities to build new connections and trust between students and members of the Ohonua community, as well as providing opportunities for the Ohonua community to explore new or different ways to build new connections with place.

In the context of physically isolated island communities in rural settings (i.e., Eua), placemaking can support community healing in the aftermath of a disaster and from the impacts of a changing climate. This can be achieved by increasing the capability of people to invest in a collective reimagining of rural and regional spaces as biophilic places that have special meaning for and with their community.

'Designing with nature in place' post-disaster is about exploring pathways for healing through nature-driven placemaking, biophilic design, or nature-based solutions. This can involve local children in tree planting to counter biodiversity loss and build place stewardship, strengthening the place capacity of a group of women or creating social opportunities for the community. This is also about bringing insights to the fore on lessons that can be exchanged and learned. City-making process has traditionally been devoid of nature discussions. This chapter offers opportunities to describe and increase awareness of the challenges but also opportunities of bringing nature into cities and communities. Pre-urbanised environments can learn from urban settings, and in return urban settings can also gain rich insights from their regional and rural counterparts.

2 Literature Review

2.1 Nature-Based Placemaking as a Design for Healing Process

Donovan, in the book 'Designing to Heal', suggests that urban design and planning are tools that can add value to humanitarian design by transforming post-disaster areas into places that open opportunities for people to heal, rebuild their lives, and thrive (Donovan, 2013). However, some of the development works implemented in the disaster aftermath fail when design is considered as a purely physical and technocratic process rather than acknowledging the complex, messy and multi-dimensional process of rebuilding (Pitzalis, 2024). In so doing, this further exacerbates disaster vulnerabilities and intensifies already existing conflicts in these places (Kammerbauer and Wamsler, 2017). A key aspect that is often ignored in the rebuilding process is recognising and listening to the multiple voices that should have been involved in the disaster aftermath (Lin, 2019). Ensuring that these voices are heard and embedded into the rebuilding process is vital to ensuring sustainable community rebuilding.

One voice that has persistently been ignored in the rebuilding process is that of nature. Evidence shows that nature can play a crucial role in the recovery process (Mabon, 2019). If given a voice, they can support the building of long-term place resilience. Amplifying nature's voice through biophilic design or enhancing ecosystem services in post-disaster recovery recognises that while disasters disrupt communities and can compound vulnerabilities, nature-based solutions provide ample opportunities for individuals and communities to heal, rebuild, and thrive.

2.2 Nature-Based Solutions to Societal Challenges, Including Disaster Risk Reduction and Recovery

Ecosystem health and human health and well-being are inextricably linked (Kiddle et al, 2021). People's mental and physical health and well-being, as well as social connections are supported by the provision of ecosystem services associated with green and blue spaces and connections with nature (Aerts et al. 2018; van den Bosch & Ode Sang 2017). In the Oceania region, these inextricable human-nature wellbeing links are embedded in the rich cultural and biological diversity of the region and the traditional ecological knowledges of the region's peoples (Kiddle et al. 2021).

With recognition of both the benefits supplied by nature and the important environmental protections provided by ecosystems, there is rising interest in nature-based solutions, as ecosystem-based approaches to address a range of challenges. Nature-based solutions are designed to address societal challenges, including climate change, water, food, and economic security; and disaster risk resilience (Cohen-Shacham et al. 2016). Nature-based solutions support the provision of benefits simultaneously for both people and biodiversity. Examples of nature-based solutions include mangrove and reef restoration—their restoration both increases coastal protection from storm surges, as well as contributes to coastal biodiversity and food security (Cohen-Shacham et al. 2016; Lovelock et al. 2024).

2.3 Understanding Socio-Ecological Health and Wellbeing

Nature-based solutions are also significant in the experience of 'sense of place', in addition to providing these environmental protections and functions (Bush et al, 2020). Connections with nature, and with specific elements of nature in place, often frame and inform our memories, our family stories, and our social networks as well as our individual emotional and mental health and place connections (Marshall et al. 2019). These place connections can be understood as 'sense of place', an umbrella term encompassing place attachments and place meanings (Masterson et al. 2019); 'sense of place' assumes 'an interconnected social and biophysical reality' (Masterton et al. 2017).

Understanding individuals' and communities' sense of place can have implications for resilience and transformative capacity (Masterson et al. 2017). Indeed, research on the impacts of disasters has repeatedly shown the significant effects on people's connection with place. Kondo et al (2024) reported that the physical destruction associated with the 2011 Japan tsunami 'disconnects people and nature'. McKinzie (2019) reported a 'negative sense of place' after significant places were destroyed by tornadoes in Missouri and Alabama, USA. Following bushfires in southern Australia, Alston et al (2018) found that 'when the places people call home are irrevocably changed by disasters, people experience a profound sense of loss—loss that includes both tangible and intangible elements'. The tangible elements include loss of homes, of infrastructure including community buildings and facilities; the 'intangible' elements include 'loss of identity, of social cohesion, of belonging and community, and health and wellbeing that are so integrally linked to one's place' (Alston et al. 2018). Alston et al. (2018) highlighted that the impacts on intangible aspects, such as the sense of place following disaster, may be of equal or greater significance than the tangible losses. This sense of loss and associated grief has been found with the destruction of ecosystems, even if there is not a direct impact on people's houses. For example, Marshall et al. (2019) recorded declining well-being, involving 'both emotional responses associated with grief, and with observable impacts on mental health' related to coral bleaching and mortality in the Great Barrier Reef ecosystem.

2.4 Post-Disaster Recovery and Healing

Recovery and rebuilding from disaster impacts require careful planning and implementation, both to ensure that local needs and aspirations are addressed, as well as to ensure that rebuilding efforts will be resilient in the face of future potential disasters, so that communities are not continually exposed to repeated disaster risk. While there is often a sense of urgency to rebuild impacted communities, their buildings and landscapes, Donovan (2012) cautions that rebuilding efforts must harness community skills and participation to ensure these contribute to individual, community, and landscape recovery.

To rebuild connections with the place following disasters, some communities are exploring cultural events and temporary urban projects as a form of 'placemaking'. Cultural events can reshape narratives and relationships to place in locations hit by disasters; cultural events can be used as an essential tool for placemaking, offering possibilities for local communities to make sense of traumatic pasts, reimagine their future, and generate a sense of place through connections between people (Fujimoto-Verdier and Martini 2024). Following the earthquakes in Ōtautahi Christchurch, Aotearoa New Zealand, temporary urban projects 'bridged the gap between emergency response and urban recovery', contributing to the reestablishment of a sense of place, as well as providing an opportunity for a strengthened engagement with Māori influence post-quake (Hobbs et al. 2022). Following the earthquake and tsunami in Fukushima, Japan, Kondo et al. (2024) noted that 'citizen-driven placemaking could potentially serve as a driving force to transform a sense of place that enhances the sense of belonging to lost places'.

3 Tonga Travelling Studio as a Case Study

The Travelling Studio program of the Faculty of Architecture, Building and Planning (ABP) at the University of Melbourne (UoM) follows a long tradition of experiential learning that offers opportunities for students to explore complex and real-life matters in unfamiliar cultures and places. The practice of uniting travel and education in design can be dated back to the early nineteenth century, when the French École des Beaux-Arts would reward the winning student of the Grand Prix competition with travel to Rome to deepen their education on classical architecture style (Middleton and Levine, 1982).

London's Architectural Association School of Architecture (AA), which was founded in 1847, is nowadays renowned for its AA Visiting School program that sets up travelling studios around the world (Architectural Association Inc., n.d.). Through collaborations and networking, some travelling studios have established long-lasting relationships, such as the MIT-THU (Massachusetts Institute of Technology—Tsinghua University) program celebrating its 30 years of existence (Blanding, 2016), and the University of Melbourne's Bower Studio approaching 20 years of design education with local and international Indigenous communities (O'Brien, n.d).

Most design students now consume information and visual experiences through technologies, yet direct experience through travel is unique, irreplaceable, and has been integrated into many Architecture and Design curricula (Kesim and Baturayoğlu Yöney, 2021). For example, at Yale University, travelling studios became policy in 2000 when funds were made available by the Henry Hart Rice Foundation (Stern and Stamp, 2016). Culver (2011) interviewed students who had participated in a travelling studio and found that this experience helped them to broaden their understanding of architecture and urban development, while gaining a sense of global awareness.

The Architecture program at the University of Melbourne has integrated post-graduate travelling studios since 1961, when the first group of students went to Japan (Fig. 1), to study its cities, buildings and culture.

Fig. 1 A group photo of students who attended Melbourne University's first-ever travelling studio Japan (No Author, 1961)

Fig. 2 A group photo of the community meeting, which explored place-based challenges and opportunities with Eua elders, youth group members, members of the Otumatafena children's group, and students from the University of Melbourne (UoM) and Tupou Tertiary Institute (TTI). *Source* Authors

The ABP Tonga travelling studio emerged from a robust partnership with the local community of Ohonua and the people's representative of Eua. The studio took place in July 2024 and ran for two weeks. It comprised of twelve ABP students, with different disciplinary backgrounds ranging from architecture to urban planning, and two studio leaders (Fig. 2). Together with the local community and local architecture students from Tupou Tertiary Institute (TTI), a local university in Nuku'alofa, the ABP students implemented a placemaking framework that explored how innovative nature-based ideas and solutions, deeply rooted in Indigenous knowledges, can assist climate change adaptation as well as individual and community wellbeing (Mateo-Babiano and Palipone, 2020b). Through consultations and activations, ABP students documented, developed, and proposed various nature-based strategies with the Ohonua community.

4 Methodology

In the next section, we adopt auto-ethnography as a lens to examine our personal experience as studio leaders of nature-based placemaking in the Tonga Travelling studio.

Ellis (2004) notes that auto-ethnography is a useful tool to systematically self-critique one's experience. The process involves critical self-reflection (graphy) of one's experience (auto) to increase one's understanding of the cultural phenomenon (ethnos-) (Ellis et al. 2011). The auto-ethnographic approach appropriately aligns with reflective research practice, taking on the role of participant observer in the implementation of the Placemaking Sandbox in a Pacific Island setting. Masterson et al. (2017) highlight the importance of 'descriptive place meanings and evaluative place attachment as tools to study the patterned variation of sense of place within or among populations' (p.48). In this case, the auto-ethnographic exercise is concerned with the extent to which nature-based placemaking can serve as a guide in Eua's post-disaster healing process.

5 Results and Discussion

5.1 The Tonga Placemaking Sandbox Experience

The travelling studio created an opportunity to build agency, new connections, and build trust between students and members of the Ohonua community, strengthening place belonging and wellbeing more broadly.

Four Teams were formed to navigate the complex cultural settings of the Ohonua community. Adopting 'nature in place' as a lens, each team adopted team names that were inspired by the local flora and fauna. The first group of ABP students called their team *Mohokoi,* which is the local name of the Ylang-ylang flower; the second group named their team *Kava*, which is a popular ceremonial drink in the South Pacific; while the third group labelled their team as *Niu* which is the local name for the endemic coconut; and the fourth group called their team *Koki*, which refers to the native parrot that can only be found on Eua island. Right from the very start, there was already a shared understanding of the role nature can play in place, bringing to the fore the importance of Eua's social-ecological system as core to the Placemaking Sandbox.

Each team engaged in Talanoa conversations with the Ohonua community to deepen their place understanding but more importantly, to build a relationship with the community, which reflects what the 'designing to heal' framework stipulates, that first, we must build trust and relationship. Accompanied by two young members of the community, teams went door-knocking around the residential blocks assigned to them. Blocks represent the spatial distribution of the township according to streets and natural topography, running mostly in parallel, from the southeast (high point) to the northwest (low point) with access to the foreshore (Fig. 3). Blocks also reflect the community structure, with churches and practitioners allocated to distinct blocks. Through intentional conversations with each household, students introduced themselves, discussing the purpose of their visit and gathering insights. This helped build trust with the community. The two accompanying local people served as translators in the process. Daily focus group discussions with various community groups were also organised in the newly built community hall (see Box 1). Walking observations were undertaken with community members, particularly with the youth and children, to capture both the natural and sensorial experiences of the place. These processes helped the student teams to work with and learn from the community about various ways to work with nature and identify nature-based opportunities. The daily community interactions also helped students build their creative, critical, and intercultural thinking competencies, and reciprocally, these processes also supported community members and groups to reimagine Eua as a biophilic place.

Box 1. Meeting with the Eua Women's Council: Reflections from the tudio Leaders Key to discovering ways out of disaster recovery and climate emergency is learning to understand the role of 'place' in shaping climate-resilient futures and building people's agency on learning about the social and ecological systems of their place, which leads to caring for the place. And if you care for the place, you look after it. The Placemaking Sandbox model is based on a place-based pedagogy that gives agency to the community to reimagine places that they care for.

Fig. 3 Aerial view of the Ohonua community township showing the coastal area on the left, which was devastated by the 2021 tsunami; Ohonua block distribution with corresponding block names (B1 - Lakataha, B2 - Toa Ko Tahakaeafe, B3 - Fele 'A Fe' Ofa' Aki, B4 - Lotolelei, B5 - Maile Lau Momo, B6 - Lolevenga, and B7 - Falemaama) *Source* Adopted from Ohonua Town Council, n.d.

Our hosts invited us, the two studio leaders, to attend the meeting of the Eua Women's Council on a sunny Friday. We arrived at the local council office where the meeting was held. Our host led the discussion. Although the studio leaders could not understand a single Tongan word in the conversation that ensued, it was fascinating to observe the animated discussions between the women in the community. They seemed to be enjoying each other's company. In between translations from Tongan to English and vice versa, what was noteworthy for us was how the discussion commenced with the re-affirmation of their community values: justice (*fakamaautotonu*), freedom (*tauataina*), sincerity (*fakamātoató*), respect (*fakaapaapa*), felicity (*fiefia*), verity (*mooni*), inclusive (*fakakau*), collaborative (*fengaueaki*), accountable (*ala fakamãua*), transparent (*ikai puli*), and flexible and adaptable (*lava ke feliliuaki*). And judging from the body language of the women, the meeting was pleasant and respectful.

The host asked us if we wanted to pose questions to the group. As community visitors, we wanted to learn more about the place and its people. We posed our first question: *'What is one thing that you love about Eua?'*.

The women paused. Before responding, the women in the community shared that no one had ever asked them this question. But readily they shared their answers. We were struck by how nature and people were embedded in their responses. They value the environment and the clean air that they breathe. They also pinpointed that their village is in one of the highest points in the whole country, which means that they are far from the ocean. Deep listening to this answer, we reflected that this individual must be expressing the ongoing trauma that is still felt by the community two years after the devastation of the 2022 tsunami.

Yet they also loved being a coastal town and close to the wharf. This shows that despite the impact of the tsunami on the community, their relationship with water is a feature that continues to define their community identity. The women in the community also identified relationships as core to the Ohonua community. They expressed that intergenerational history, marriage, and community relationships were things that they love about their community.

The second question we posed was: *'What change do you want to see in your community/place?'* Environmental concerns were front of mind in their responses. First, they worried about the ongoing soil erosion and coastal erosion that had been happening on the island, particularly during monsoon season when the community would experience heavy rains. Compelling evidence already exists that these extreme events are becoming more frequent and severe. There is already extensive documentation of the direct impacts of climate change, particularly on island community settings such as Eua. These include coastal soil erosion, wetland loss, and escalating risks of natural hazards such as hurricanes/typhoons, storm surges, and flash flooding. While not as obvious are the indirect effects of increasing levels of poverty and biodiversity loss.

The need to respond and address the intensifying impacts becomes urgent and imperative (Dasgupta et al. 2014).

There were also two other critical strands of change they wanted to see. They wanted to improve community skills and the standard of education. These included improvements to Eua's basic infrastructure provision, such as road improvement, and to improve town cleanliness (addressing the roaming pigs) and the beautification of the Eua landscape, particularly in relocation areas.

They also expressed the need to work together. This stems from the highly segregated approach the different blocks take in conducting tasks. Different blocks have different religious affiliations, and the residential blocks tend to work together with their religious brothers and sisters, yet there is only one place that they want to improve.

5.2 Placemaking Projects and Stories of Place

Building social and human capital in island communities could contribute to addressing health inequities (McIntosh et al. 2019), exclusion of specific subpopulations (Stanley et al. 2019), and cultural isolation (Townsend 2008). Social infrastructure refers to facilities and services that improve the quality of life of the community and contribute to their health and well-being (Infrastructure Australia, 2019).

The Placemaking Sandbox methodology 'is the creative way of co-designing space into meaningful places' (p. 5), which can be adopted in a studio, fostering an experimental and experiential environment for creative problem solving (Mateo-Babiano and Palipane 2020a).

Four groups were formed with a focus on developing a placemaking framework for the Ohonua community. Teams also co-designed place-based ideas to enhance social capital, which is underpinned by a deep realisation of the criticality of nature as core to human survival. As earlier mentioned, each group commenced by first adopting team names that were inspired by the local flora and fauna.

Through a nature-based placemaking process, student teams in partnership with community members came up with ideas and ways to help strengthen place belonging and reduce community vulnerabilities. They co-designed nature-based opportunities and implemented four tactical placemaking initiatives that demonstrated biophilic design, nature-based, participatory design approaches to shape nature-centred public spaces, revealing important insights on the potential role of nature-driven placemaking in building back better.

The first group, Team Mohokoi, focused on Blocks 1 and 2 (the foreshore area of Ohonua), as the setting for their placemaking strategy. This was an area that was hard hit by the tsunami of 2022, hence was converted into a no-build zone. Working and interacting closely with the Otumatafena planting group, a group of children from

Blocks 1 and 2 who re-planted the foreshore area with trees, the Mohokoi team, through storytelling discovered the children's strong bond with and deep connection to place by transforming the area from a place of trauma to a place for caring. The children continued to care for the local trees they planted, learning more about fostering healthy habitats. They provided an excellent example of building place and nature stewardship, demonstrating both resilience and care for place, empowering the community in embedding nature-based solutions in the development of Lakataha or the foreshore area.

The second group, Team Kava, focused on Blocks 3 and 4. Their name was inspired by the kava plant, which is the traditional Polynesian beverage of the same name. The group explored the flora native to Eua Island. From their investigation, students discovered the 'Ohai' tree. They adopted the tree as their team's project mascot. With one 'Ohai' tree located between Blocks 3 and 4, students used this tree to strengthen the relationship between people and nature in Eua through cultural heritage, history, and nature elements that can be found only on the island. Based on their historical research, they found that 'Ohai' was highly valuable as it was the traditional material to build boats. When living on an island, fishing boats are vital to their everyday life and livelihoods. These provide access to their livelihoods, for communication, transportation, and inter-island trading. Nature celebration, which enabled communities to thrive on Eua, was the base of their placemaking idea.

The third group, Team Niu, focused on Blocks 5 and 6, building their placemaking project on the stories shared by community members, particularly one elder who inspired the community, and the team, as she shared many stories and brought hope to everyone. One particular story left an indelible mark on the team. This is the story of the Ovava tree.

She shared that when Tonga was severely affected by the tsunami in early 2022, the Ovava tree planted on block 5 was one of the very few trees that were left standing. She also shared a story about the same tree 50 years prior. The Ovava tree played an important role in the community's place attachment. It served as a social space and a place to mind their children as they wait for the arrival of the ferry from Nuku'alofa, the main island and Tonga's capital.

Reflecting on their engagement with the community, the team learned many precious stories. Indeed, they quickly realised that the community is the expert. Several key lessons derived from their encounter. As placemakers, they learned that one should have a fresh and open mind, especially in accepting different ideas from the community. Gathered from informal conversations, the group focused not on finding the solutions, but to first build trust and relationship, and gain a deeper understanding of the local knowledge through the sharing of stories. They did not make assumptions.

By doing so, they were able to generate a solution that better catered for the needs of the community. And through their reciprocal storytelling with the community, the team felt the people's strong connection to their surrounding and their very strong love of place.

Team Niu's placemaking idea was inspired by the idea of waiting under the Ovava tree. As a tactical exercise, the team engaged the children in the block to create

Fig. 4 Gathered under the large, leafy Ovava tree, children and adults of the Ohonua community co-design and co-develop the Ovava playground, a placemaking activity led by Team Niu (Lee, 2024)

the 'Ovava Playground' project by using scrap natural materials that can be found around the community. In the end, they were inspired by adopting nature-based placemaking as a catalyst to foster sense of belonging, restore connections with nature post-disaster, and address urgent climate issues through recycling and waste management, promoting resilience for future generations of Eua (Fig. 4).

The fourth group, Team Koki, focused on blocks 7 and 8. Block 8, located further inland and higher on the hilltop, is the relocation site for households from Blocks 1 and 2 who have been moved to this area as they lost their homes in the foreshore because of the tsunami. On their first day, two young residents accompanied each student team to show them the blocks. During the reflection session, students found the stories confronting as they share about their unimaginable loss of connection to their homes, ancestors and familial spaces. At the same time, they also expressed their desire to find community connection, to re-connect with the ocean and move past the traumatic event. While the sharing of stories of joy and trauma felt overwhelming, the teams truly appreciated the meaningful sharing and deep listening to the perspectives of the two, and the broader community. During the community meeting that was organised a few days later, the team continued to feel the tension of the community and the trauma of being relocated. It was highly perceptible when talking to this group of the Ohonua community. Beyond losing a place, a home near the sea, near churches and faith groups, these people have also lost their gardens. The community expressed their strong connection to their gardens. Gardens are essential to their food supply and the focus of daily activities in Ohonua. Being relocated away from the township and on the edge of the forest presents challenges in maintaining a garden and food supply while wild pigs are roaming around. Beyond brick-and-mortar solutions to protect gardens, students have patiently listened to people, first informally and then with a more structured consultation, giving space for all groups, men and women, young and old, to express their ideas.

During this event, an elderly woman explained that there was a simple solution to keep the pigs away: a fast-growing native plant that is spiky and can be planted in a tight grid. This knowledge of a traditional, cost-effective, and ecosystem-based approach to protect crops from wild pigs was about to disappear until a space was given for some group of the community to express themselves and to be heard.

Thereafter, Team Koki focused on shaping a women-led leadership initiative, which was concerned with empowering locally embedded leadership to create a resilient and united Ohonua grounded in full community representation, working towards collaborative governance. This reinforces the contributions of nature-based placemaking to adaptive governance.

6 Reflections and Critical Lessons for Cities and Urban Communities

What lessons do pre-urbanised environments in small island community settings offer to urban regions? How can the lessons of Eua's tsunami responses and nature placemaking inform urban disaster resilience, and in turn, what can Eua learn from urban-based experiences? In this section, we explore the ideas of island urbanism, particularly in the Oceania context. Island urbanism is one way of understanding and representing the unique and diverse ways of life of those residing in and visiting Pacific Island communities, such as Eua (in Tonga).

Oceania, which includes Pacific Island communities, covers almost a third of the Earth's surface, making it a significant context to learn from. With over 1300 unique languages that remain active and spoken throughout the islands, the region appears to be fragmented physically and culturally, with diverse socio-cultural representations. Yet these small island communities are linked biogeographically, with flora and fauna as well as climate and geology interlinked, we can learn about how these factors can influence but also be influenced by urbanisation processes in the context of climate change.

6.1 Urbanisation is a Spectrum

Urban areas in Oceania started as island communities or coastal settlements. Through the urbanisation process, these settlements experienced large-scale transformation.

Accompanying urban development are the multiple challenges the cities and communities face due to processes of global warming and climate change, and the local implications that follow.

Many of these island communities are experiencing structural change. This change may be economic, social or spatial. This can be due to increased mobility or influenced by the growing importance of the service sectors, which continue to transform communities, some to become post-industrial knowledge-based economies. (Indeed, in Oceania, many urban areas 'leapfrogged' the industrial developmental stages partially or entirely.) Yet a key insight from this project is that fostering innovation does not require high-tech solutions. Through placemaking, more resilient communities are shaped. We describe these communities as those that are aware of the richness that is present within their communities. In the case of Pacific Island communities such as Eua, this is about incorporating the nature lens as central to community shaping and highlighting nature features as core (complementing the people-focused approach in placemaking) to the healing process. Across Oceania, there are diverse examples of how nature and nature-based solutions are being adopted both to address climate change and other challenges, as well as strengthen community belonging and inclusion.

Incorporating 'biodiversity-sensitive urban design' strategies provides important lessons to urban conurbations, lessons that can be learned from island communities. Moreover, these communities are already aware of and depend on nature, from food resources to the presence of towns that are teeming with pigs, dogs, and chickens in public areas.

6.2 Surfacing Diverse Voices in Placemaking

Healing from disasters, as highlighted by Donovan (2013), is a complex and messy process, but must be underpinned by recognising and listening to multiple voices. In the aftermath of Eua's tsunami, many voices were neglected. The placemaking process placed importance on seeking out and listening to these diverse voices—the student voice, the community voice and Nature voice. The studio process was committed to creating stronger collaborative partnerships with local students and institutions as well as the local community, which served as a key strength that guided the Placemaking Sandbox.

6.3 Students as Partners

The travelling studio is an educational offering that is uniquely place-based, where learning purposely goes beyond the classroom. In the placemaking studio, the Ohonua community became the setting for intentional intercultural interactions. In such a setting, Melbourne University students encountered people different from them but also similar. Coming from diverse, social backgrounds yet engaging with community as diverse as them, these interactions, which are seldom captured in classroom instructions, provided students with opportunities to engage with communities and the local government to gain skills and understanding of the processes that are necessary to create places from inception through to implementation, which is core to placemaking pedagogy. In this instance, students became community partners to deliver a place-based endeavour that helped create agency and enhance the placemaking capacity of Ohonua adults and children, inspiring them to invest urban spaces with positive place meaning (Mateo-Babiano and Palipane 2020a).

6.4 Community as Experts

The key to sustainability and stewardship is building relationships through stronger partnerships and purposeful collaborations, which is vital to purposeful growth. In doing so, it allows learners to collaborate with community members to tackle complex 'big issues' such as waste management, biodiversity, and climate change. Through place storytelling and truth-telling (in the four different student projects), placemaking ideas helped create awareness about nature perspectives; it increased community agency, and through deep engagement with local communities and the co-production of place-based design ideation responses. It was also an opportunity for the visiting students to learn Indigenous practices such as making baskets and coconut leaf balls.

6.5 Nature-Based Placemaking

Nature is critical to our survival at all scales: in our local places and at a global scale. Centring nature-in-place in the travelling studio's post-disaster community building endeavours offered lessons on how to integrate nature's voice into place, an approach to valuing biodiversity in placemaking practice. This provided the opportunity for community and students to learn together and learn from each other, with implications for improving individual and community health and well-being. For example, by identifying the stories of significant trees, the students and young people fostered and renewed intergenerational connections to place and created new opportunities for new stories to be shared about the trees. And by listening to diverse voices, old ways of harnessing nature's services (to control damage from pigs) were uncovered and reinvigorated.

7 Conclusion and Implications

This paper responds to the call to improve the education of future built environment professionals (e.g., planners, urban designers, and placemakers) by building their place-based knowledge as they engage with Pacific Island communities, particularly in post-disaster settings. In the end, learners valued the alternative classroom experience and non-traditional learning methods of placemaking, acknowledging the importance of building intercultural competencies that helped them to navigate the complex cultural settings of Eua.

This is to ensure that we as professionals facilitate change by bringing to the fore the multiple voices, particularly the voices of the children, the women, and other community members, in creating a socially and culturally safe space; to connect those who are learning how to shape better communities with various community members who can provide authentic and constructive inputs to their place.

Placemaking has become central to the relationship-building process. And it can also serve as a capacity building strategy to support community climate action by increasing the capability of people to invest in a collective reimagining that can transform rural and regional spaces into nature places for their community.

Nature-based placemaking offers the potential to guide and ground community rebuilding and the healing process, embedding nature at its core. While this acknowledges the importance of understanding and maintaining positive human–nature interactions, it is also vital to not romanticise such processes and acknowledge potential community vulnerability, particularly to typhoons and tsunamis.

Yet placemaking is a process that builds the capacity for people to have agency over (or with) the systems of which they are a part. In healing communities and rebuilding places, it can assist in highlighting the critical role that nature can play in shaping a more holistic socio-ecological system. In island communities, it is about building opportunities to re-learn traditional ecological knowledges, connection to place and stewardship of place (Kingsley et al. 2009; Pedersen Zari et al. 2019). And in so doing, it shapes the agency of the social and ecological systems, leading to caring for and loving these places so that nature, people, and place can harmoniously thrive.

Acknowledgements Funding: Judy Bush is the recipient of an Australian Research Council Discovery Early Career Researcher Award DE240100699 (2024–2027).

We acknowledge the financial and administrative support of the Faculty of Architecture Building and Planning, especially Katie Frank, Aftab Hussain and Ruth Woodgate, in delivering the Tonga Travelling Studio.

We acknowledge the support of our hosts, Hon. Taniela Fumisimalohi, Dr. Caroline Tupoulahi-Fusimalohi and Dr. Ungatea Kata, as well as the precious support of students enrolled in the Tonga Travelling studio, students and those from the Tupou Tertiary Institute and the members of the Ohonua community.

References

Aerts R, Honnay O, Van Nieuwenhuyse A (2018) Biodiversity and human health: mechanisms and evidence of the positive health effects of diversity in nature and green spaces. Br Med Bull 127(1):5–22. https://doi.org/10.1093/bmb/ldy021

Alston M, Hargreaves D, Hazeleger T (2018) Postdisaster social work: reflections on the nature of place and loss. Aust Soc Work 71(4):405–416. https://doi.org/10.1080/0312407X.2017.1409776

Architectural Association Inc (n.d.) AA school. AA School. https://www.aaschool.ac.uk/

Blanding M (2016) "Beijing studio" celebrates 30 years. MIT News | Massachusetts Institute of Technology. https://news.mit.edu/2016/beijing-studio-celebrates-30-years-0714

Bush J, Hernández-Santin C, Hes D (2020) Nature in place: placemaking in the biosphere. In: Hes D, Hernández-Santin C (eds) Placemaking fundamentals for the built environment. Palgrave Macmillan, pp 39–61

Cohen-Shacham E, Walters G, Janzen C, Maginnis S (2016) Nature-based solutions to address global societal challenges. International Union for Conservation of Nature. www.iucn.org/resources/publications

Culver LD (2011) The influence of study and travel abroad on the personal and professional development of students in architecture design programs. https://doi.org/10.25148/etd.fi11042601

Dasgupta P, Morton JF, Dodman D, Karapinar B, Meza F, Rivera-Ferre MG, Toure Sarr A, Vincent KE (2014) Rural areas. Climate change 2014: impacts, adaptation, and vulnerability. In: Field CB, Barros VR, Dokken DJ, Mach KJ, Mastrandrea MD, Bilir TE, Chatterjee M, Ebi KL, Estrada YO, Genova RC, Girma B, Kissel ES, Levy AN, MacCracken S, Mastrandrea PR, White LL (eds) Part A: global and sectoral aspects. Contribution of working group II to the fifth assessment report of the intergovernmental panel on climate change. Cambridge University Press, Cambridge, United Kingdom and New York, NY, USA, pp 613–657

Donovan J (2012) Designing to heal; reconstructing communities after disasters. Proc Inst Civ Eng-Urban Des Plan 165(2):69–77. https://doi.org/10.1680/udap.10.00009

Donovan J (2013) Designing to heal: post-Disaster rebuilding to assist community recovery Part B: designing a process and product for recovery. Environ Des Guide (78):1–9

Ellis C (2004) The ethnographic I: a methodological novel about autoethnography. Rowman Altamira.

Ellis C, Adams TE, Bochner AP (2011) Autoethnography: an overview. Historical social research/ Historische sozialforschung, 273–290

Fujimoto-Verdier M, Martini A (2024) Healing from trauma in post-disaster places?: Placemaking, machizukuri and the role of cultural events in post-disaster recovery. In: Trauma informed placemaking, pp 253–262. https://doi.org/10.4324/9781003371533-26

Hobbs M, Ahuriri-Driscoll A, Kingham S, Wiki J, Marek L, Dionisio MR, Curl A, Schluter P, Banwell K, Mackenbach JD (2022) A city profile of Ōtautahi Christchurch. Cities, 121, Article 103481. https://doi.org/10.1016/j.cities.2021.103481

Kammerbauer M, Wamsler C (2017) Social inequality and marginalisation in post-disaster recovery: challenging the consensus? Int J Disaster Risk Reduct 24:411–418

Kesim B, Baturayoğlu N (2021) Architectural travelers: the role of field trips in spatial design education. Period Polytech Arch 52(2):155–164. https://doi.org/10.3311/ppar.18861

Kiddle GL, Pedersen Zari M, Blaschke P, Chanse V, Kiddle R (2021) An oceania urban design agenda linking ecosystem services, nature-based solutions. Tradit Ecol Knowl Wellbeing Sustain 13(22):12660. https://doi.org/10.3390/su132212660

Kingsley J, Townsend M, Phillips R, Aldous D (2009) "If the land is healthy … it makes the people healthy": the relationship between caring for Country and health for the Yorta Yorta Nation, Boonwurrung and Bangerang Tribes. Health Place 15(1):291–299. https://doi.org/10.1016/j.healthplace.2008.05.009

Kondo T, Ghezelloo Y, Sakaguchi N (2024) Emerging place governance to lost places for communities: government planning and citizen-driven placemaking in the disaster memorial parks after the 2011 Japan tsunami. Int J Disaster Risk Reduct 114, Article 104907. https://doi.org/10.1016/j.ijdrr.2024.104907

Lee A (2024) Figure 4. Children playing under the Ovava tree.

Lin PSS (2019) Building resilience through ecosystem restoration and community participation: post-disaster recovery in coastal island communities. Int J Disaster Risk Reduct 39:101249

Lovelock CE, Bennion V, de Oliveira M, Hagger V, Hill JW, Kwan V, Pearse AL, Rossini RA, Twomey AJ (2024) Mangrove ecology guiding the use of mangroves as nature-based solutions. J Ecol 112(11):2510–2521. https://doi.org/10.1111/1365-2745.14383

Mabon L (2019) Enhancing post-disaster resilience by 'building back greener': evaluating the contribution of nature-based solutions to recovery planning in Futaba County, Fukushima Prefecture, Japan. Landsc Urban Plan 187:105–118

Marshall N, Adger WN, Benham C, Brown K, I Curnock M, Gurney GG, Thiault L (2019) Reef grief: investigating the relationship between place meanings and place change on the Great Barrier Reef Australia. Sustain Sci 14(3):579–587

Mateo-Babiano I, Palipane K (eds) (2020a) *Placemaking Sandbox*. Palgrave Pivot. https://doi.org/10.1007/978-981-15-2752-4.

Mateo-Babiano I, Palipane K (2020b) Reimagining place through the sandbox studio pedagogy: an introduction. In: Placemaking sandbox: emergent approaches, techniques and practices to create more thriving places, pp 1–14. Springer Singapore, Singapore

Masterson VA, Enqvist JP, Stedman RC, Tengö M (2019) Sense of place in social–ecological systems: from theory to empirics. Sustain Sci 14(3):555–564. https://doi.org/10.1007/s11625-019-00695-8

Masterson VA, Stedman RC, Enqvist J, Tengö M, Giusti M, Wahl D, Svedin U (2017) The contribution of sense of place to social-ecological systems research: a review and research agenda. Ecol Soc 22(1):Article 49. https://doi.org/10.5751/ES-08872-220149

McKinzie AE (2019) You don't miss it 'til it's gone: insecurity, place, and the social construction of the environment. Environ Sociol 5(3):232–242. https://doi.org/10.1080/23251042.2019.1574463

McIntosh K, Kenny A, Masood M, Dickson-Swift V (2019) Social inclusion as a tool to improve rural health. Aust J Primary Health 25(2):137–145

Middleton R, Levine N (1982) The beaux-arts: and nineteenth-century French architecture. Thames and Hudson.

Author, 1961No Author (1961) First ABP travelling studio, Japan. The University of Melbourne Archives.

O'Brien D (n.d.) Bower Studio. https://bowerstudio.msd.unimelb.edu.au/

Pedersen Zari M, Kiddle GL, Blaschke P, Gawler S, Loubser D (2019) Utilising nature-based solutions to increase resilience in Pacific Ocean cities. Ecosyst Serv 38:100968. https://doi.org/10.1016/j.ecoser.2019.100968

Pitzalis S (2024). Expertise versus aspiration: ethnography of post-disaster reconstruction in Emilia. In: Disasters and changes in society and politics. Bristol University Press, pp 155–170

Stern RA, Stamp J (2016) Pedagogy and place: 100 years of architecture education at Yale. Yale University Press

Townsend R (2008) Adult education, social inclusion and cultural diversity in regional communities. Aust J Adult Learn 48(1):71–92

Tupoulahi-Fusimalohi. (n.d.) Figure 3. Ohonua township with block distribution

United Nations Environmental Programme (UNEP) (2021) Making peace with nature: a scientific blueprint to tackle the climate, biodiversity and pollution emergencies. https://www.unep.org/resources/making-peace-nature.

Van den Bosch M, Ode Sang Å (2017) Urban natural environments as nature-based solutions for improved public health – a systematic review of reviews. Environ Res 158:373–384. https://doi.org/10.1016/j.envres.2017.05.040

"Socialise with Friends or Family while Enjoying the Scenery": Biophilic Approach to Youth Social Engagement

Parisa Ziaesaeidi and Tracy Washington

Abstract Previous studies demonstrate parks are primary places for active and passive recreation and can provide opportunities for young people to participate in activities that include socialising with others. The inclusion of natural elements such as trees, green areas, and water features could enrich the experiences of visitors, encourage a range of recreational activities and strengthen community bonds. Within this context, there is a need to better understand how nature-based park settings can foster social engagement within neighbourhoods. This study will explore different forms of youth social engagement, with a specific focus on activities occurring within natural features of neighbourhood parks, to address the research question: how does the biophilic approach to park design influence youth social engagement?

This study informs a regenerative approach to urban and regional planning by introducing a qualitative approach in terms of social engagement. Within this context, the data were collected by using a youth-friendly visual method (photo elicitation approach) with 192 youth aged 9–17 years from the Moreton Bay Region (MBR) of South-East Queensland, Australia. This research found that parks with trees, rocks, gardens, wooded areas, and water bodies allow the engagement in diverse forms of passive and active social activities such as sitting, talking, and playing. Based on these findings, we suggest the creation of small natural areas, known as "pocket forests," in parks with trees, rocks, gardens, wooded areas, and water bodies as a potential way to engage youth socially within their neighbourhood spaces. This chapter presents strategies to assist decision-makers with planning more green urban settings as a nature-based solution to foster more youth social engagement.

Keywords Biophilia · Green urban settings · Social engagement · Youth · Nature-based approach

P. Ziaesaeidi (✉)
School of Engineering, Design and Built Environment, Western Sydney University, Sydney, Australia
e-mail: p.ziaesaeidi@westernsydney.edu.au

T. Washington
School of Architecture and Built Environment, Queensland University of Technology, Brisbane, Australia

A. Cheshmehzangi et al. (eds.), *Nature in Cities, Nurturing Cities*, Urban Sustainability,
https://doi.org/10.1007/978-981-95-5487-4_10

1 Introduction

Providing access to neighbourhood facilities, such as parks, is a common component included in theories and principles about good urban design (Bush 2020). Parks within neighbourhoods perform essential functions to meet people's needs such as contact with nature in an urban landscape (Lee et al. 2015; Matsuoka and Kaplan 2008). They also provide a place for social interaction that creates a sense of community (Goličnik and Ward Thompson 2010; Kazmierczak 2013).

Neighbourhood parks are often primarily designed to include activities for children, emphasising playground areas (Zhou et al. 2022). These spaces frequently fail to meet the needs of young people, who are rarely involved in the planning process. As a consequence, youths above 13 yrs. old are often drawn to city centres and shopping areas for social interaction (Woolley and Johns 2001). However, appropriately designed parks, or park features can play a crucial role in fostering the social well-being of youth, impacting their satisfaction, quality of life, and the sustainability of communities (Kazmierczak 2013; Peters et al. 2010; Ziaesaeidi et al. 2020; Ziaesaeidi 2025).

Research on youth social engagement in parks, as highlighted by Kazmierczak (2013) and Peters et al. (2010), often suggests that providing green spaces and natural features creates more opportunities for social interaction and help to promote social cohesion among youth. Conversely, reductions in the number of green spaces and natural features in cities can limit possibilities for youths' outdoor activities (Kruiz et al. 2019). Natural features in parks not only support diverse recreational activities but also contribute to positive mental well-being, enriching the social fabric of communities, as evidenced by extensive research (e.g., Wolf et al. 2015; Peters et al. 2010; Maas et al. 2009; Kazmierczak 2013). These green spaces offer opportunities for social interaction, often providing a context to learn important social skills, such as cooperation and collective decision-making (Ziaesaeidi et al. 2020). These types of social activities significantly contribute to the socio-cultural education and overall development of youth (Ibid, 2020).

Following this, there is a need to explore what types of natural features within parks best facilitate youth social activities, providing enhanced opportunities for social engagement. Many studies have explored integrating biophilic design elements into landscapes to enhance people's health and well-being (Zhong et al. 2022; Ryan et al. 2014; Clancy and Ryan 2015). Parks, serving as direct connections to nature, offer numerous opportunities for the integration of biophilic design elements (Ryan et al. 2014). Biophilia in park design involves incorporating elements inspired by nature, such as trees and water bodies, to enhance the sensory experience of visitors (Idib, 2014).

By integrating biophilic design elements such as trees and water bodies, parks create a sensory experience that promotes relaxation, stress reduction (Heerwagen and Hase 2001), and overall social engagement. Creating an environment with biophilic design beneficial to social engagement means establishing a setting that encourages people to interact, connect, and participate in various activities (Konijnendijk et al. 2013; Heerwagen 2009). When individuals feel relaxed and connected to the natural surroundings, it becomes easier for them to engage with others, whether through casual conversations, shared activities, or community events. Fostering an overall environment conducive to social engagement in this context means designing and maintaining park settings that facilitate and enhance social interactions among park visitors. This chapter explores how biophilic approach in park design impacts social engagement among youth. To do so, this research will answer this question:

RQ-How the biophilic approach in park design can influence youth social engagement?

Data were collected using a photo elicitation method then coded and thematically analysed. Activities raised by youth regarding each photo were categorised as either active or passive forms of youth social activities. This chapter indicates how nature-based park settings can influence youth social engagement. To the authors knowledge, it is the first time that biophilic approach in park design has been investigated in relation to social engagement among youth.

2 Background and Definitions

Park design significantly influences social interactions (Kazmierczak 2013). With ongoing urbanisation, understanding the impact of incorporating biophilic principles into park design on social engagement is essential. This section of chapter provides a comprehensive outline of key concepts related to social engagement and the biophilia approach in park design.

2.1 Social Engagement

Pescosolido et al. (2007) defines social engagement as "the extent to which an individual participates in a broad range of social activities". Social activities are shaped by, and embedded in, the environment in which they take place (Clark and Uzzell 2002; Michler et al., 2o19; Kariyawasam et al. 2020) and are an important aspect of the broader social environment (McNeill et al. 2006). Socioecological studies show that activities of park users can be influenced by factors including various social environments (Mahdiar and Dali 2016).

Social activities can be described as passive or active forms of interaction with others (Jenks 1998). Passive social activity occurs when park users share the same space without any or with little direct interaction. Active social activity occurs when park visitors have both verbal and visual encounters (Simões Aelbrecht 2016).

Passive social activity not only provides the opportunity to coexist among others but also enables individuals to see and hear others within the shared park setting (Gehl 1987). Therefore, through passive forms of social activities, being alone can provide opportunities for the youth to have an acquaintance with an unknown person through sharing the same park settings and participating in activities happening therein (Ziaesaeidi et al. 2023). Just knowing that others are present, hearing them and witnessing their activities can contribute to a sense of community and social connectedness (Ziaesaeidi et al. 2023). However, being alone in parks without others being present cannot provide a feeling of community and connection. Activities such as solo play within sight of others (Mohammadi Tahroodi and Ujang 2022) could be considered as passive social activity.

Active social activities occur when park users have both verbal and visual encounters including conversations, meeting friends/new people (Trombeta and Cox 2022). Active social activities are characterised by the multifaceted nature of human connections, where individuals not only coexist within a shared environment but actively participate in verbal and visual encounters (Ibid., 2022). This participation can involve engaging in conversations and activities, thereby fostering connections with both familiar faces and new acquaintances, and contributing to the vibrant social fabric of the community (Ziaesaeidi et al. 2023). In the context of parks, individuals actively participate in the parks, transforming these public spaces into dynamic hubs. Therefore, people contribute to a collective dynamic through lively conversations and visual interactions. This interconnected web of social activities plays a crucial role in shaping parks environments' vibrancy, inclusivity, and overall well-being, turning parks into centres of community engagement and shared experiences (Trombeta and Cox 2022).

2.2 *Theory of Biophilia*

The biophilia hypothesis suggests that humans possess a biologically based attraction to certain aspects of the natural environment and that their well-being depends, to a great extent, on the relationships with the surrounding natural world (Kellert 1997; Kellert 2002; Kellert 2008; Ulrich 1993; Wilson 1994; Wilson 1984). Fromm (1973) proposed the possibility that the deep affiliations humans have with nature are rooted in our biology. Unlike phobias, which are the aversions and fears that people have of things in the natural world, philias are the attractions and positive feelings that people have towards certain habitats, activities, and objects in their natural surroundings (Jones 2013).

In the realm of parks and urban planning, a biophilic approach seeks to strengthen the bond between individuals and the natural environment within the built landscape (Clancy and Ryan 2015). By incorporating elements like natural landscapes and creating green urban settings, application of this concept encourages people to connect with nature. This approach aligns with a nature-based perspective in urban planning (Colléony and Shwartz 2019).

Biophilic design integrates natural elements into architectural and landscape designs, fostering a connection between people and their environment for improved well-being. Limited research has explored the role of biophilia in park design. For example, Hady's (2021) study introduced a sustainable landscape approach using biophilic design patterns. It revealed that elements like filtered sunlight, specific plantings, water features, natural textures, and nature views positively impact mental well-being, community engagement, and overall environmental satisfaction. The social benefits of the biophilic approach in park design refer to the positive impacts on social interactions and community engagement. This approach emphasises creating living environments within parks that foster a strong connection between individuals, promoting a sense of community, relaxation, and overall social well-being. However, there is a lack of research on the social benefits of the biophilic approach in park design. This chapter explores how the biophilic approach in park design can influence youth social engagement based on literature and the author's research.

2.3 Biophilic Urban Acupuncture (BUA)

Urban acupuncture is a socio-environmental theory that parallels traditional Chinese acupuncture by applying small-scale, targeted interventions to relieve stress in the urban fabric, much like how acupuncture relieves tension in the human body (Lerner 2003). These interventions act as socially catalytic nodes within the city and aim to produce meaningful positive change without large-scale disruption.

Biophilic Urban Acupuncture (BUA) introduces a "nature-based" dimension by integrating biophilic design principles into these interventions. BUA emphasises the deliberate placement of small biophilic elements, such as green spaces, natural materials, and water features (Ojo 2024), strategically woven through urban threads and nodes to improve mental health, foster connections to place, and enhance overall well-being.

What distinguishes BUA from other urban design approaches, such as tactical urbanism or generic small-scale interventions like pocket parks, is its explicit focus on the ecological and psychological benefits rooted in biophilia. Tactical urbanism often prioritises social or spatial restoration of urban environments. BUA aims to directly benefit ecological systems by creating habitat connectivity and enhancing biodiversity within the urban environment. Simultaneously, it supports social cohesion and mental health by providing residents with regular, accessible exposure to nature (Ojo 2024). Biophilic Urban Acupuncture as a transformative approach that integrates ecological restoration with social and spatial activation, contributing uniquely to sustainable urbanism.

The effectiveness of BUA is particularly pronounced in dense urban areas, where pedestrian mobility ensures consistent daily contact with these biophilic nodes. In contrast, suburban environments, with auto-centric design and sprawling land use, offer fewer opportunities for such regular engagement. This highlights the importance of context in applying BUA principles.

3 Approaches to Data Collection from Youth

To better understand youth social engagement through a biophilic approach in park design, photo elicitation was used in this research. Photo elicitation offers a powerful addition to standard data collection and presentation techniques. Introducing photographs to questionnaire settings can allow participants to both tell and show their stories. The emotions that are brought forth from interaction with images during the survey can aid in producing richer data for analysis. Using photo elicitation is also a useful tool when conducting research with youth, since selecting images can garner reactions, emotional responses, and meanings that may not have been accessible using verbal or text-based methods alone (Copes et al. 2018).

The data collected and presented in this chapter formed a component of the first author's PhD research and included an original selection of 32 photos. Each photo contained a dominant park feature (e.g., a playground, a pathway, an open green space, a dog off-leash area, a stream, a skate park, a basketball court, or a community garden). The current study aims to explore how a biophilic approach in park design impacts youth social engagement. The first author selected the photos (out of 32), each featuring a prominent natural element like gardens, wooded areas, or water bodies (Kaczynski et al. 2008). Examples of the selected 32 photos include open playing fields, off-leash dog areas, streams, or community gardens. This selection aims to improve clarity and facilitate analysis. The participants completed the photo elicitation tool within approximately 10 to 15 minutes via iPad. Ethics approval was obtained with approval number 2000000474. A total of 198 participants (youth aged 9–17) were approached in eight different neighbourhood parks and two youth community centres within the Moreton Bay Region (MBR), South-East Queensland, Australia. Data collection took place across 22 days (six weekdays and 16 weekend days), from 2:00 pm to 5:00 pm. Youth were presented with sets of four photos and tasked with choosing their preferred one. Subsequently, they engaged in discussions about the activities depicted in the chosen photo (see Fig. 1).

Table 1 shows the balance between male and female participants and demographic information of participants.

Fig. 1 Sample of one of the participant's responses to the questions regarding selected photo (*Source*: the authors)

Table 1 Demographic information of participants

# of youth participants	# of invalid responses	% of Females	% of Males	% of total age range of participants
N = 198	6	50.79%	49.21%	9 (8.47%) 10 (17.46%) 11 (14.82%) 12 (14.29%) 13 (15.34%) 14 (10.58%) 15 (6.35%) 16 (7.41%) 17 (5.29%)

Source the authors

The collected data consists of both visual (their top choice of the 4 photos) and textual information (activities preference comments). These were analysed separately through descriptive statistics using Microsoft Excel-2010). Data were explored through inductive thematic analysis. The first author categorised the activities preferences for each photo setting raised by participants and then combined the codes into themes, shown in Table 4. After describing potential themes and their possible definitions, the first author re-organised code families into theme's codes: passive and active social activities.

For the analysis, photos were categorised based on the dominant features in two groups: hardscape (defined as constructed using natural and man-made materials, such as stone or concrete walkways, decks, walls, planter boxes and anything else that uses hard materials); and softscape (defined as living elements, such as trees, grass, vines and shrubs, water elements and areas such as woodlands and wetlands) (see Table 2).

As it was possible to see both soft and hardscapes in one photo, the first author calculated the percentage of each element within each photo using AutoCAD-2019 and Photoshop-2019. As shown in Table 2, the indicated hardscape and softscape percentages implies that the entire image is not entirely one or the other. Therefore, it is important to note that this computation does not encompass elements like the sky and background features, such as play equipment, which contribute to the overall visual appearance but are not categorised within hardscape or softscape percentages.

The intent was to consider the photos that represented a biophilia lens within varied park scenes. From the pool of 32 photos, 22 photos had a greater percentage of soft scape and included semi-natural/natural spaces, open spaces and predominantly natural features. The selected photos included elements such as logs, tree stumps, boulders, plants, drainage paths, rocks, sand, and walkways or bridges connected to creeks and naturally shaded areas (see Table 3).

Then a concept map was created to determine the relationship between codes and codes' families (See Fig. 2). This helped to organise the diverse social activities articulated by youth into structured classifications, and to identify which activities

Table 2 The categorisation of photos portraying softscape and hardscape

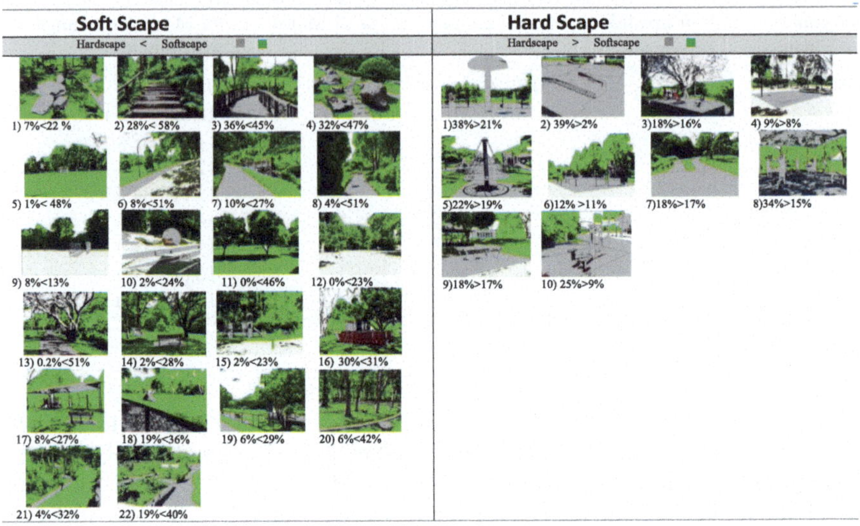

Source the authors
[*] *The categorisation is based on the dominant features in each photo-photo credits (all 32 photos): ARC Linkage Project LP160101341.*

were more likely to happen in which park setting. The social activities described by youth were divided into active and passive themes:

Active: types of activities during which youth have both verbal and visual interactions with other people, including direct conversations, and meeting friends/new people.

Passive: types of activities through which youth share the same space without any, or with little direct interaction with other people.

4 Findings

The findings of this research show that natural areas such as wetlands, bushland, and wooded areas provide opportunities for both active and passive forms of social engagement. In their responses, youth talked about both forms of social activities. The main active social activities occur when youth have both verbal and visual encounters including "hanging out", "sitting and talking", and "picnicking", and passive social activities which occur when youth share the same space without any or with little direct interaction including "playing" and "relaxing". Photos numbers 3, 6, 7, and 14 (see Table 4) illustrate that natural features such as water and a boardwalk and naturally shaded areas situated within grassed areas with scattered trees could provide time for youth to not only spend time with their family and friends but also alone.

Table 3 22 photos and the dominant natural features within each photo

 #1 Stream or creek in a wooded area with rocks connected creek	 #2 Rainforest pathway with trees, bushes	 #3 Pathway with elements such as water and a boardwalk	 #4 Manicured pathway scene with rocks, shrubs, and green grass	 #5 Large open space with trees and green grass
 #6 Naturally shaded areas and manicured park with green grass and scattered trees	 #7 Pathway with green grass with scattered trees and shrubs	 #8 Manicured pathway scene with shrubs and green grass	 #9 Manicured amphitheatre with green grass	 #10 Wooded area within green grass
 #11 Green turf grass and scattered trees within a large open area	 #12 Manicured park with green grass and scattered trees	 #13 Manicured park with green grass and scattered trees	 #14 Wooded and naturally shaded areas	 #15 Community garden with a mown grass pathway
 #16 Manicured pathway with green grass	 #17 Dog off-leash area with trees	 #18 Green turf grass and scattered trees with play equipment	 #19 Scattered trees with play equipment	 #20 Community garden with a gravel pathway
	 #21 Large open space with trees and green grass with a cricket net	 #22 Large open space with trees and play equipment		

Source the authors

Additionally, these natural features can provide opportunities for youth to have little direct interaction. This can be the case when going to parks alone or in company.

According to Table 4, large open and naturally shaded areas, and manicured parks situated within grassed areas with scattered trees provide opportunities to interact with others and meet new people. Photos with planters or community gardens were described by participants as providing more learning opportunities and features that allow youth to be connected to nature. Photos with open green spaces could provide more opportunities for ball games and teamwork/group activities. These natural elements provide opportunities for youth to play, explore, imagine, and be challenged by natural features. Natural elements could enable youth to move freely around the environment allowing them to explore, run, jump, climb, crawl, feel, smell, and more. The results also revealed that youth are more likely to interact with

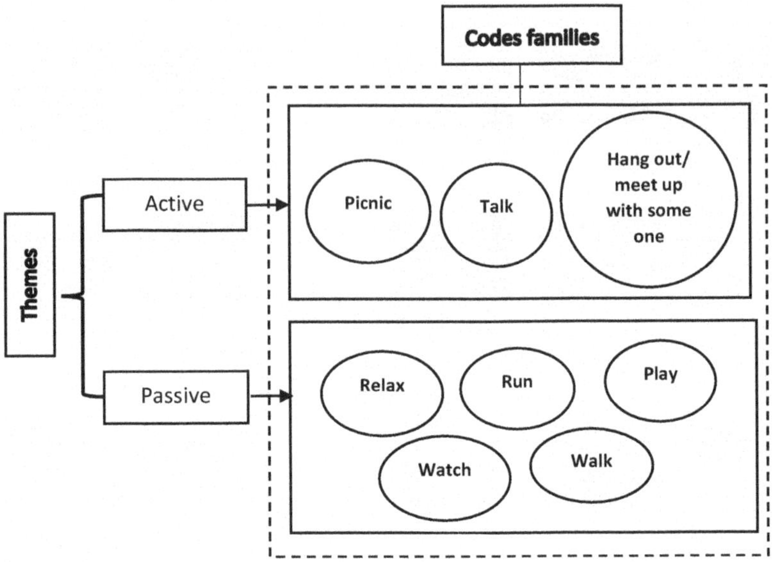

Fig. 2 Categorisation of activities (*Source*: the authors)

nature in the context of fun activities, suggesting an association between nature and these fun or relaxing moments.

These results, taken together, suggest the importance of nature in providing a context for social interactions and activities, places where fun and relaxing activities can contribute to youth social engagement. A preference for natural elements as a context for fun and social activities supports the biophilia theory. This implies that humans not only associate nature with emotional happiness but also possibly a desire to experience nature because of the social experiences and relaxation it facilitates. This finding also points to the potential synergistic effect of social activities occurring in the presence of nature. Different from the other contexts analysed, fun activities are likely to be a social setting where youth tend to interact with each other in a group.

In addition, the authors undertook further analyses to better understand which natural features are preferred for either active forms of social engagement or passive forms of social engagement (see Tables 4). Table 4 illustrates themes raised by participants and codes for each theme. According to Table 4, "codes" refer to specific themes that represent different types of activities. Each code represents a particular kind of activity, either classified as "Active" or "Passive". The listed examples under each category serve as instances or instances of these broader activity types. For example, "Hang out/meet up" and "Sit and talk" are specific codes under the "Active" category, while activities like "Watch/explore the wildlife" and "Relaxing" fall under the "Passive" category. These codes help organise and categorise data for analysis and interpretation in this research. The frequency of each theme helps identify which natural features could provide opportunities for either passive or active social activities.

The selected features (Tables 3 and 4), particularly #1 (stream or creek in wooded area), #2 (rainforest pathway), and #3 (pathway with boardwalk and water elements) represent high-impact, small-scale interventions that align well with the core principles of Nature-Based Urban Acupuncture. Their scale creates intimate, sensory-rich environments that support both passive and active youth engagement through relaxation, exploration, and subtle social interaction. These features function as "nodes" within a broader ecological network, enhancing connectivity by drawing young users through natural corridors that link diverse park areas. Similarly, features like #6, #7, and #8, with manicured paths, scattered trees, and shaded areas, provide restorative pockets that are easy to access and promote casual, everyday encounters, thereby amplifying social cohesion. Larger open areas such as #11, #12, #13, #21, and #22 act as scalable acupuncture points, offering flexible zones for group games and spontaneous activities, while still integrating natural elements that ground them in biophilic design. By focusing on localised, nature-based interventions that are socially catalytic and ecologically sensitive, these features exemplify how BUA principles can be applied to transform everyday public spaces into vibrant, inclusive, and resilient environments for youth.

5 Biophilia Concept and Active/Passive Social Engagement

Referring to Table 4, the dominant finding demonstrates that natural features such as manicured amphitheatre with green grass (#9) are mostly preferred by youth for active forms of social activities and engagement such as chatting and talking, having lunch and hanging out with family and friends. As a boy aged 13 yrs. old pointed out, he can "sit down and relax with my friends and brother" or as a girl aged 12 mentioned "I think it would be a good place to hang about with my friends". Another girl aged 10 yrs. old said she could "sit on the steps and have a picnic" and a girl aged 10 yrs. old said she could "run, walk and play with friends".

According to the detailed analysis of the findings (see Tables 3 and 4), other natural features such as manicured pathway scene with rocks, shrubs (#4), naturally shaded and manicured areas situated within green grass with scattered trees (#6 and 12), and the wooded area within green grass (#10) are also preferred by youth for both active and passive social engagement. Youth talked about hanging around, talking, and picnicking for the photos with these features. These mentioned activities are likely to provide opportunities for directly connecting youth with their friends and families or others. Their comments are included in Table 4.

A number of naturalistic features listed below provide suitable contexts for passive forms of social engagement such as playing, relaxing and enjoying the view, having

Table 4 The identified themes and its codes based on each photo

Photos	Frequency of youth activities	Themes	Codes	What would you like to do in this scene?
 #1	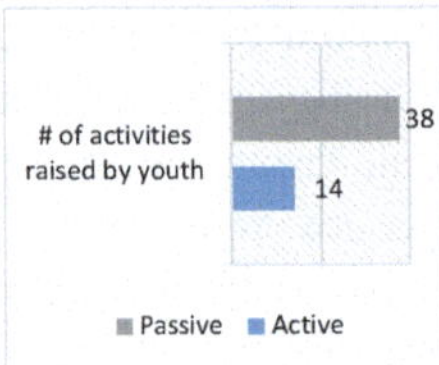	Active	"Hang out" "Meet up" "Sit and talk" "Picnic"	"Nice scenery. photos, meet up with friends" a girl 14 yrs old "Sit and talk" a boy 12 yrs old
		Passive	"Watch/explore the wildlife" "Relaxing" "Play" "Jumping over the rocks" "Walking" "Swimming" "Riding a bike"	"Have an adventure with my friends" a girl 13 yrs old "I would walk with my family and dog" a girl 10 yrs old
 #2	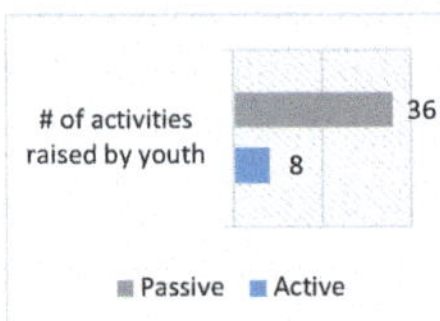	Active	"Hang out" "Talk" "Picnic with family/friend(s)"	"Meet up with friends and family" a girl 14 yrs old "Talk while walking with friends/parents" a girl 12 yrs old "Have time out in nature, I really enjoy this activity with family" a girl 12 yrs old
		Passive	"Relaxing" "Look/watch at view/ explore" "Bush walk" "Climbing"	"Enjoy the nature and tell my parent's what I've learnt in school about nature" a girl 9 yrs old "I would like to walk there and admire the cool plants and I would see many cool animals because it looks like lots of birds would live there" a girl 10 yrs old
 #3	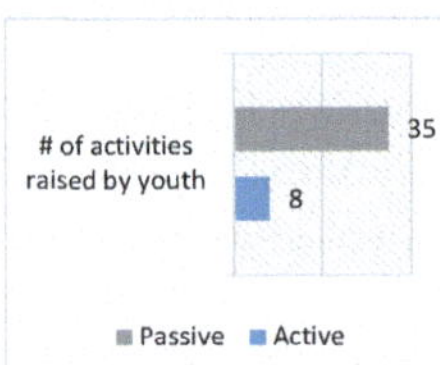	Active	"Walk and talk"	"This will be a fun place to go for walks with your family and other families and cousins etc it is a family place" a boy 13 yrs old "Socialise with friends or family while enjoying the scenery" a girl 16 yrs old
		Passive	"Look/watch at view"; "Fishing"; "Riding"; "Walk"	"Most likely take walks with my friends or pets" a boy 17 yrs old

(continued)

Table 4 (continued)

 #4	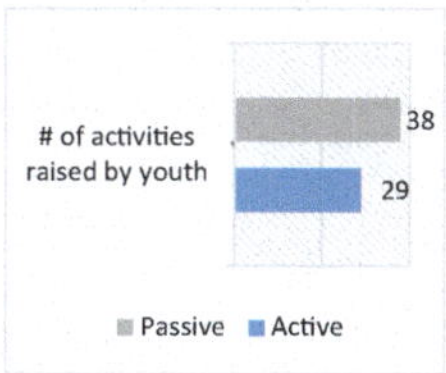	Active	"Hang out" "To do activities with friends and family" "Picnic" "Talk"	"Have a picnic and have quality family time" a girl 12 yrs old "Enjoy the scenery with my family or friends and socialise with them" a girl 16 yrs old "I would walk and talk with my great auntie" a girl 10 yrs old
		Passive	"Relaxing" "Social walk" "Climbing on rocks and trees"	"Sit down and relax, eat, have picnics, climb on the rocks" a girl 14 yrd old
 #5	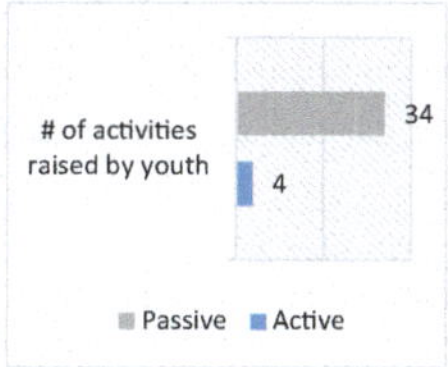	Active	- "Hang out"	"Walk around and talk" a boy 10 yrs old
		Passive	- "Play" - "Walk/walk the dog" - "Running around alone/ with my animal"	"Again, fitness and playing around with friends and family" a girl 13 yrs old "Run around with my animals" a girl 13 yrs old
 #6	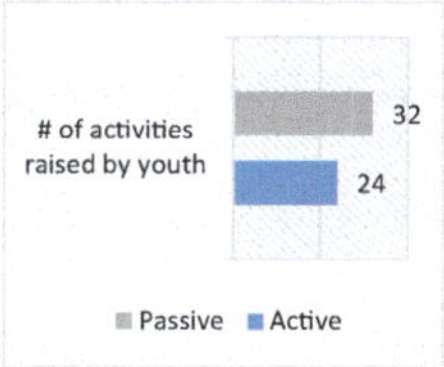	Active	"Hang out" "Picnic" "Chat"	"Have conversations with my friends" a girl 16 yrs old
		Passive	"Sit and relax" "Play"	"Sit down and enjoy nature" a girl 12 yrs old "Chill and relax with others" a girl 9 yrs old "Play with mates and take my dog for walks" a boy 11 yrs old
 #7	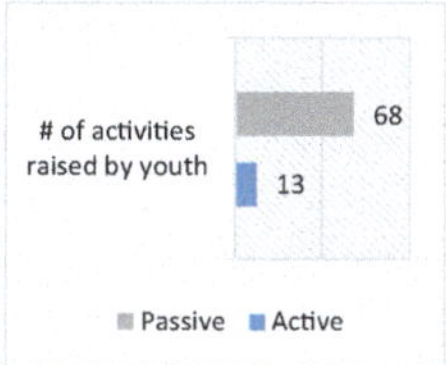	Active	"Hang out" "Walk and talk"	"I would walk around and chat with whoever is with me" a girl 12 yrs old
		Passive	"Walk/walk the dog" "Ride bikes"	"Explore the nature and scenery with people" a girl 13 yrs old "Walk with family and friends" a boy 13 yrs old

(continued)

Table 4 (continued)

 #8	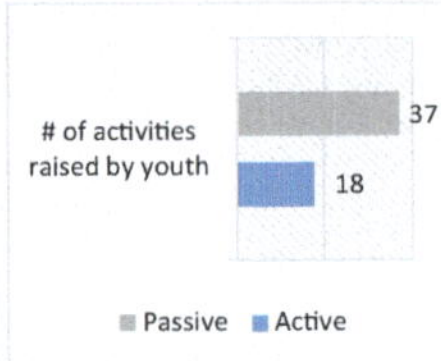	Active	"Walk and talk/talk" "Picnic"	"Talk to friends while doing runs and walks" a boy 13 yrs old
		Passive	- "Relaxing" - "Walk" - "Watch/look the scenery"	"I would relax and enjoy the view with my family and friends" a girl 13 yrs old "Nature suitable activities!!" a girl 13 yrs old
 #9	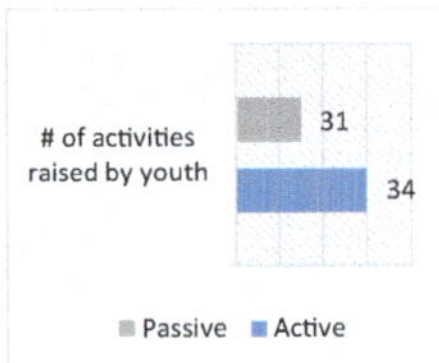	Active	"Hang out" "Chat and talk" "Picnic"	"Have a picnic I and a fresh air around the area with family and friends" a boy aged 13 yrs old "Have conversations with my friends" a girl aged 16 yrs old
		Passive	"Relaxing" "Chill and enjoy the view" "Running" "Climbing the ball" "Play"	"Climb on the big giant ball" a girl aged 10 yrs old
 #10	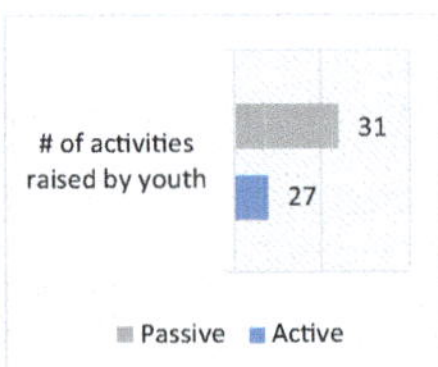	Active	"Hang out" "Picnic"	"Sit and enjoy food with friends and family" a girl 13 yrs old
		Passive	"Relaxing" "Explore" "Play" "Running around"	"Race on the filled play tag and have dinner there with friends and family" a boy 11 yrs old
 #11	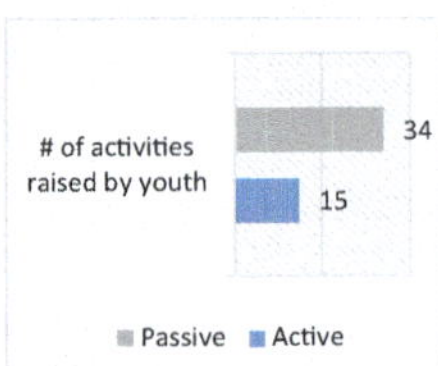	Active	"Hang out"	"Like I said there is a range of activities here other than soccer it will be a fun place to go to with multiple people and the only thing missing here is some play area with actual fun stuff in it and a basketball court" a boy 13 yrs old
		Passive	- "Play ball games" - "Ride a bike" - "Run"	"I would love to play soccer with my friends" a boy 11 yrs old

(continued)

Table 4 (continued)

 #12	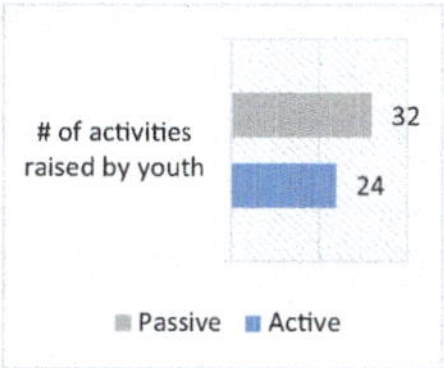	Active	"Picnic"; "Enjoy the time with my parents"	"Enjoy time with my parents" a boy 12 yrs old
		Passive	"Relaxing" "Walk" "Daily exercise"	"I would take photos and relax with my family and/or friends" a girl 13 yrs old "I would want to climb the tree if I am allowed and I would chat with my friend/sister" a girl 12 yrs old
 #13	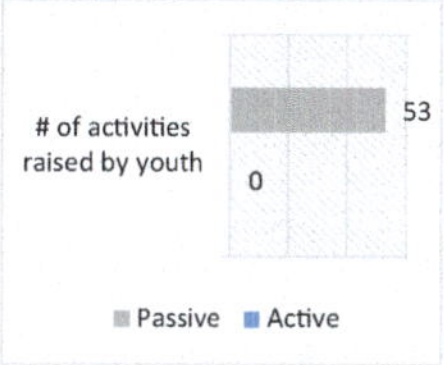	Active	———	———
		Passive	"Walk/run the dog" "Play"	"Take my dog to socialise with other dogs" a girl 16 yrs old "Work out and enjoy" a boy 10 yrs old "I would like to sit down and just be outside" a girl 12 yrs old
 #14	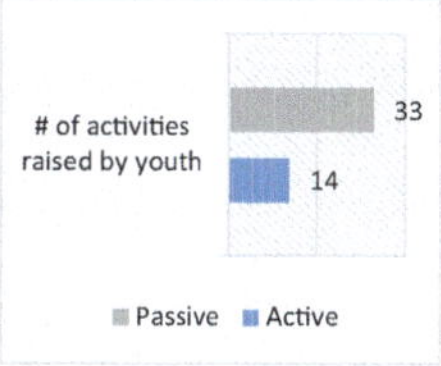	Active	"Hang out"	"Hang out with my friends" a girl aged 14 yrs old
		Passive	"Play" "Relaxing" "Climbing trees"	"I can play hide and seek" a boy 10 yrs old "Playing chase and climbing" a girl 9 yrs old "It looks like a good place to walk or sit and clear my head" a girl 13 yrs old
 #15	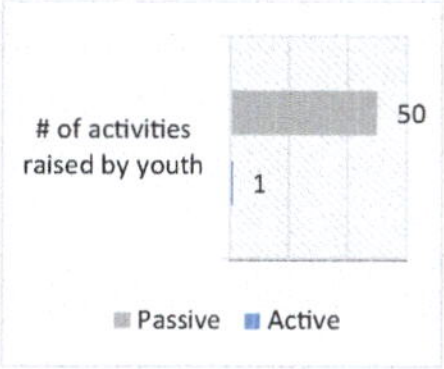	Active	"Picnic"	"Walks, picnics, observe plants, insects and birds that live there" a girl 16 yrs old
		Passive	- "Watering the garden" - "Watch the plants"	"Be able to help with watering the garden" a girl 10 yrs old "Walk around with relatives and friends" a boy 11 yrs old "I think having a public garden would be great for insects and birds, as well for people to enjoy the herbs and plants growing there" a girl 16 yrs old

(continued)

Table 4 (continued)

 #16	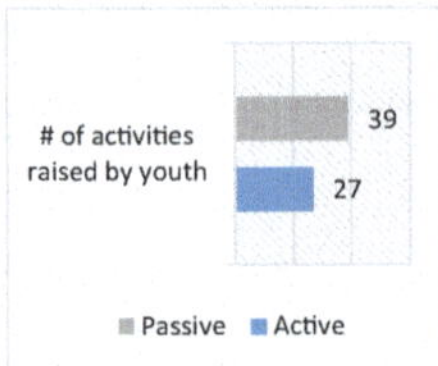	Active	"Picnic"; "Talk"; "Hang out"	"Having picnics, celebrate my birthday and play lots of games with my friends and family" a girl aged 10 yrs old
		Passive	"Walk" "Play" "Running" "Skateboard" "Watching" "Ride a bike"	"I like to look at animals that may be around such as birds" a girl aged 12 yrs old "At these parks I would like to enjoy the time spent with my friends and enjoy the scenic view whilst we walk" a girl aged 12 yrs old "Walk, explore nature" a girl aged 17 yrs old
 #17	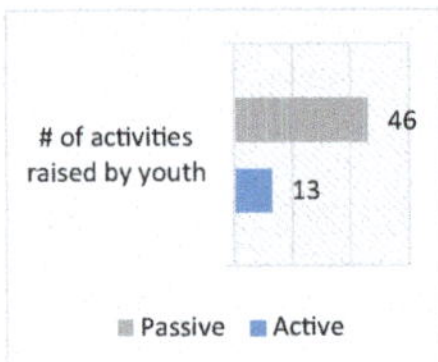	Active	"Talk"	"Play games and socialise" a boy aged 11 yrs old
		Passive	"Play/ play with dog" "Running around/ running with my dog" "Walk/ walk with my dog"	"Run around with my pets" a boy aged 11 yrs old "To play tag" a boy aged 9 yrs old
 #18	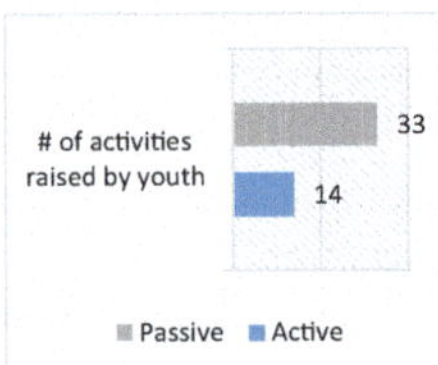	Active	"Hang out"; "Talk"	"Hang out with friends, let younger children enjoying being outside and even fitness" a girl aged 13 yrs old
		Passive	"Play" "Running around" "Climbing"	"Socialise with friends while climbing the equipment" a girl aged 16 yrs old "Play with my friends" a boy aged 10 yrs old
 #19	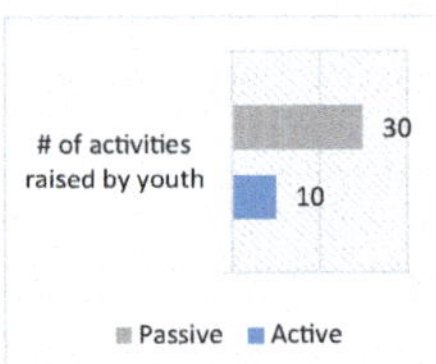	Active	"Spending time with my siblings"	"I would like to spend time with my siblings" a girls aged 12 yrs old
		Passive	"Play" "Running around" "Sit and watch"	"Bring my niece to enjoy outside other than social media" a girl aged 17 yrs old "Play with my dog" a boy aged boy 15 yrs old

(continued)

Table 4 (continued)

 #20	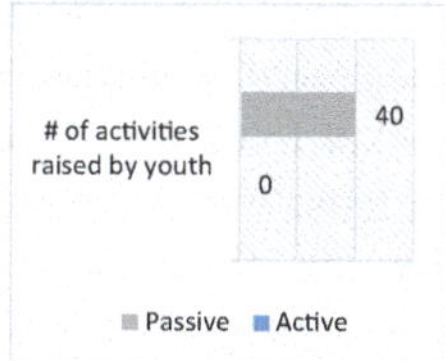	Active	———————	———————
		Passive	"Participating in planting" "Watch the plants" "Walk"	"To take part in growing vegetables and fruits, and learn to source things locally" a girl aged 15 yrs old "Walk around the plants" a girl aged 11 yrs old
 #21	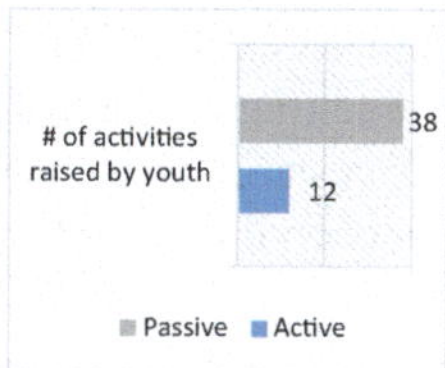	Active	"Picnic"	"Play a friendly game of footy or relax and have a picnic with a few friends" a girl aged 15 yrs old
		Passive	"Play" "Walk/walk the dog" "Running around with my animal"	"Play soccer and other sports with my friends" a boy aged 16 yrs old "Play cricket with my mates" a boy aged 17 yrs old
 #22	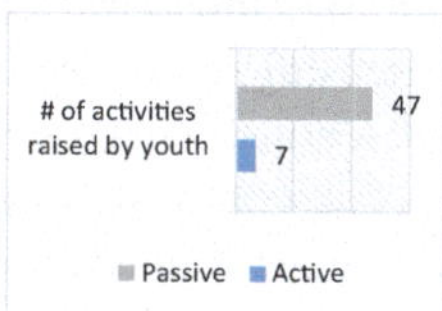	Active	"Hang out"	"Have some fun with my friends" a girl aged 13 yrs old
		Passive	- "Play" - "Relaxing"	"Sit and relax and play" a boy aged 12 yrs old "Just play around" a boy aged 10 yrs old

Source the authors
Note while the table categorises youth activities under "Passive" and "Active" social engagement, it is important to acknowledge the frequent overlaps in how youth described their experiences. Many statements suggest that a single natural feature can support both. These overlapping themes highlight the fluid and dynamic nature of youth engagement

the adventure with siblings/friend and family. These features can offer youth opportunities to participate in group activities, which in turn foster social connections and interactions with different people.

- a stream or creek in a wooded area with rocks connected creek (#1),
- a rainforest pathway with trees, and bushes (#2),
- a pathway with elements such as water and a boardwalk (#3),
- a large open space area with trees and green grass (#5 and #11 and #21),
- naturally shaded areas(#8),
- manicured park situated within green grass with scattered trees (#14),
- pathway with elements such as green grass with scattered trees (#7),
- manicured pathway scene with shrubs and green grass (#13),
- a community garden with a manicured pathway (#20).

According to the findings, firstly, as youth are more engaged through passive forms of social engagement, the park settings should offer opportunities for more social activities and a variety of unplanned and planned social uses by providing diverse natural features to offer more social connections. Nature-based settings with more elements such as a large open space situated within green grass with scattered trees are likely to actively engage youth. Choosing a pocket forest as a nature-based setting is deliberate due to its unique potential to engage the younger demographic. While larger open spaces within green landscapes with scattered trees can be appealing, smaller, more concentrated natural settings like pocket forests have proven advantages. These settings, comprising trees, rocks, gardens, wooded areas, and water bodies, have shown promise in actively involving youth. By creating these pocket forests within neighbourhoods, they offer diverse opportunities for social engagement among youth. The intimate scale and varied elements in a pocket forest provide a more accessible and manageable space for youth to explore and interact with nature, fostering a sense of community. This type of setting can activate neighbourhoods more socially as it offers a localised, easily accessible area for various activities and connections, ultimately contributing to a more vibrant and interconnected community.

This "nature-based" urban environment such as pocket forest fundamentally extends the traditional concept of urban acupuncture by embedding ecological and biophilic elements into the urban fabric, rather than focusing solely on social or spatial revitalisation. They intentionally integrate natural features that provide ongoing psychological, social, and environmental benefits. For youth, this means that the targeted interventions are not just catalysts for urban renewal but become meaningful places that foster deeper connections with nature and community. As demonstrated by youth preferences for natural features like manicured amphitheatres, shaded pathways, and pocket forests, these biophilic settings encourage both passive and active social engagement, promote mental well-being, and support a sense of belonging. Thus, BUA's transformative potential lies in its ability to create multifunctional nodes that simultaneously enhance ecological systems, support social cohesion, and enrich everyday urban experiences, a triple benefit that sets it apart from other small-scale urban interventions.

This dual focus on social and ecological revitalisation distinguishes BUA as a transformative approach that not only activates neighbourhoods socially but also restores and strengthens urban ecosystems, offering youth enriched environments for both play and relaxation.

6 Discussion and Implications

The sustainable development of urban areas not only shapes the environmental landscape but also influences the intricate social dynamics present in city life (Turner-Skoff & Cavender 2019). Neighbourhood parks are recognised as pivotal elements that enhance social interactions and foster community cohesion (Barrera et al. 2016).

Within this context, this study examined the concept of biophilia in parks design in relation to youth social engagement by using a unique method, photo elicitation. The main finding of this research is that these informal urban green settings can act as a powerful design tool for socially re-connecting neighbourhoods with micro scale nature and thus inspiring sustainable behaviours such as spending quality family time.

Kuo et al. (1998) found that the more trees and greenery that form part of inner-city spaces such as neighbourhoods, the more these spaces can be used by residents of different ages. In addition, their study found that, compared with residents living near barren spaces, those closer to greenery enjoy more social activities, have more visitors, know more of their neighbours, and have stronger feelings of belonging. Our main findings are in accordance with this showing that natural elements such as a large open space situated within green grass with scattered trees are likely to actively engage youth within neighbourhoods by providing more opportunities for gathering places for youth, their families and friends to communicate with each other. In addition, natural elements such as trees, rocks, gardens, wooded areas, and water bodies can encourage diverse forms of social activities such as youth becoming acquainted with others through fun activities.

Well thought out strategies should provide a context for the development of neighbourhood with social affordances that allow community social engagement and build upon the social activities and engagements prevailing in the communities. Our findings show this could happen by designing and building more natural and green settings distributed within neighbourhoods. Parks with waterways, bushland, tree canopy and green ground cover, and open spaces, when strategically planned, designed, and managed, support a good quality of social life in an urban environment by providing opportunities for diverse social engagements. As youth were found to be mostly socially engaged through spending time with their family, green infrastructure such as parks within the neighbourhoods that support family interaction and passive interaction with others, support sustainable communities. Our findings show that natural elements in parks are key elements making local neighbourhoods more welcoming and socially inclusive by connecting diverse people.

In addition, the authors focused on introducing and evaluating the "pocket forest" concept to enhance environmental benefits while creating opportunities for community engagement and well-being. This concept involves developing small, forested areas in urban spaces, such as parks, integrating natural features like trees, plants, water bodies, and wildlife. By integrating more natural elements in urban areas including neighbourhood parks, this approach not only helps address pressing environmental issues such as the urban heat island effect but also significantly contributes to connecting young individuals with their peers, families, and fellow residents within neighbourhoods. This multifaceted approach aims to create spaces that not only promote environmental sustainability but also encourage vibrant social connections and a sense of community. Incorporating more green elements into urban areas not only will help balance temperatures and reduce the urban heat island effect (Priyadarsini 2009) but also provide opportunities for connecting youth with their friends and families or other neighbourhood residents.

6.1 Practical Implementation

Previously, urban and strategic planners and landscape architects focused on nature-based solutions and bioclimatic principles in urban green designs as an action to the climate change adaptation (Jones et al. 2022). The role of social engagement in our urban and strategic planning as a successful approach for community engagement lacks concentrated attention. This study presented that natural elements such as large open space situated within green grass with scattered trees, rocks, gardens, wooded areas, and water bodies can encourage diverse forms of social activities within neighbourhoods. According to its findings, the design of natural elements in parks could help with engaging youth within our urban areas like neighbourhoods.

These natural elements not only foster well-being and play but also enhance connectivity by linking everyday routes and habitats where youth can engage with nature and others. This is in accordance with the connectivity principle of urban acupuncture that emphasises how small but strategically placed interventions can catalyse broader ecological and social benefits.

Given space constraints, the findings highlight the importance of nature-based urban acupuncture sites that exemplify the principles of scale, precision, connectivity, and impact.

First, the manicured amphitheatre with green grass (#9, see Table 3) serves as a well-defined social intervention. This offers a precisely scaled space where youth actively engage in chatting, picnicking, and relaxing. Its formal design, combined with green elements, creates an inviting micro-environment that supports high social connectivity and interaction. This setting enhances neighbourhood vibrancy by facilitating frequent, informal gatherings that support peer bonding and emotional well-being.

Second, the wooded pocket forest (#10, see Table 3) illustrates how a concentrated natural area can function as a transformative ecological intervention. With elements such as trees, rocks, and shaded grass areas, it provides a multi-sensory environment that supports both active and passive social engagement. Youth described this space as ideal for "talking," "playing," and "picnicking," reflecting a desire for semi-enclosed, flexible environments that allow for both solitude and sociability. Its small, manageable scale and natural complexity foster a deep sense of place and ecological intimacy, exemplifying how BUA integrates environmental and social function.

These examples demonstrate how small-scale, strategically located interventions can transform underutilised urban areas into socially vibrant and ecologically resilient spaces. Their intimate scale and diverse biophilic elements create immersive, manageable environments that resonate with youth. Together, they function as "needles" that relieve urban stress, improve mental well-being, and foster a sense of belonging.

It is important for urban planners and landscape architects to consider including biophilic elements in urban areas to impact social well-being in addition to other benefits. Natural elements can not only act as the nature-based solutions to physical problems such as pollution and the urban heat island effect in urban green designs but can also support opportunities for diverse social engagements. In order to inform planners, landscape architects, and other designers, this study proposes a biophilic approach to park design, one that seeks to understand diverse forms of social engagement within neighbourhood green spaces.

6.2 Gaps and Future Studies

This study found the lack of attention to how the full spectrum of biophilic elements in park design contribute to social engagement. Further research should therefore examine a wider sample of green urban settings and their impact on social engagement. This should include natural features of various scales and types and their qualitative impact on social engagement be assessed. This research introduced a qualitative approach, i.e., photo elicitation for collecting data from youth. The authors acknowledge that there would be some bias in the data since much of the data was collected using provided photographs of various park settings (researcher-generated). Therefore, there is a need for future research to ask youth to take photos of their favourite green urban settings not limited to provided photographs. More research should also be undertaken on the development of pocket forest in developed urban contexts with scarce availability of spaces. Finally, more studies are needed to identify and create best practice solutions relating to the provision of natural settings that facilitate social interaction.

7 Concluding Remarks

This chapter affirms the prominent role of green urban settings such as neighbourhood parks in facilitating different forms of social engagement. By applying small-scale, precisely located interventions, such as pockets of woodland, shaded pathways, water-integrated trails, and open green areas, urban designers can catalyse broader social, as well as environmental benefits.

This study demonstrates that socially focused development of green spaces such as Nature-Based Urban Acupuncture (BUA) can serve as powerful tools for ecological transformation and youth social engagement. Embedding small-scale, biophilic interventions into urban green spaces can make a big difference. The findings reveal that youth are drawn to natural features, such as shaded areas, water elements, and open grassy spaces, that support both passive and active forms of social interaction. Unlike conventional tactical urbanism, BUA emphasises both social vitality and ecological enhancement, promoting biodiversity, mental well-being, and community connection. These insights support the need for planners and designers to prioritise biophilic principles in urban green space design, ensuring inclusive, engaging, and ecologically resilient environments for young people.

Acknowledgements The authors would like to express their sincere gratitude to all the young participants who actively participated in this research. Their willingness to share their insights have been invaluable. Their participation has played a significant role in advancing the goals of this research. We appreciate their commitment to making a positive impact and fostering a sustainable future.

References

Barrera F, Reyes-Paecke S, Harris J, Bascuñán D, Farías JM (2016) People's perception influences on the use of green spaces in socioeconomically differentiated neighborhoods. UFUG 20:254–264

Bush J (2020) The role of local government greening policies in the transition towards nature-based cities. Environ Innov Soc Transits 35:35–44

Chang CC, Cheng GJY, Nghiem TPL, Song XP, Oh RRY, Richards DR, Carrasco LR (2020) Social media, nature, and life satisfaction: global evidence of the biophilia hypothesis. Sci Rep 10(1):4125

Choe Y, Schuett MA (2020) Stakeholders' perceptions of social and environmental changes affecting Everglades National Park in South Florida. Environ Dev 35:100524

Cirianni FMM, Comi A, Luongo AS (2022) A sustainable approach for planning of urban pedestrian routes and footpaths in a pandemic scenario. TeMA 15(1):125–140

Clancy J, Ryan C (2015) The role of biophilic design in landscape architecture for health and Well-being. Landsc Arch Front 3(1):54–62

Clark C, Uzzell DL (2002) The affordances of the home, neighbourhood, school and town Centre for adolescents. J Environ Psychol 22(1–2):95–108

Clark, C, Uzzell, D L (2006). The socio-environmental affordances of adolescents' environments. *Children and their environments: Learning, using and designing spaces*, 176

Colléony A, Shwartz A (2019) Beyond assuming co-benefits in nature-based solutions: a human-centered approach to optimize social and ecological outcomes for advancing sustainable urban planning. Sustainability 11(18):4924

Copes H, Tchoula W, Brookman F, Ragland J (2018) Photo-elicitation interviews with vulnerable populations: practical and ethical considerations. Deviant Behav 39(4):475–494

Dannenberg AL, Jackson RJ, Frumkin H, Schieber RA, Pratt M, Kochtitzky C, Tilson H (2003) The impact of community design and land-use choices on public health. A Sci Res Agenda 93(9):1500–1508

Fromm, E (1973), The Anatomy of Human Destructiveness, Holt, Rinehart and Winston, New York

Gehl, J (1987). Life between Buildings: Using Public Space, [trans by Jo Koch, Van Nostrand] Reinhold: New York

Goličnik B, Ward Thompson C (2010) Emerging relationships between design and use of urban park spaces. Landsc Urban Plan 94(1):38–53. https://doi.org/10.1016/j.landurbplan.2009.07.016

Hady SIMA (2021) Activating biophilic design patterns as a sustainable landscape approach. J Eng Appl Sci 68(1):1–16

Han M, Kim M (2018) A critical review of child-friendly environments, focusing on children's experiential perspectives on the physical world for sustainability. Sustainability 10(10):3725

Heerwagen J (2009) Biophilia, health, and Well-being. Restorative commons. Creat Health Well-Being Urban Landscs:39–57

Heerwagen J, Hase B (2001) Building biophilia: connecting people to nature in building design. Environ Des Constr 3:30–36

Jenks C (1998) Active/passive. Core Sociological Dichotomies, 261–274

Jones DR (2013) 'The Biophilic university': a de-familiarizing organizational metaphor for ecological sustainability? J Clean Prod 48:148–165

Jones L, Anderson S, Læssøe J, Banzhaf E, Jensen A, Bird DN, Zandersen M (2022) A typology for urban green infrastructure to guide multifunctional planning of nature-based solutions. Nat-Based Solut 2:100041

Kaczynski AT, Potwarka LR, Saelens BE (2008) Association of park size, distance, and features with physical activity in neighborhood parks. Am J Public Health 98(8):1451–1456. https://doi.org/10.2105/AJPH.2007.129064

Kariyawasam S, Wilson C, Rathnayaka LIM, Sooriyagoda KG, Managi S (2020) Conservation versus socio-economic sustainability: a case study of the Udawalawe National Park. Sri Lanka Environ Dev 35:100517

Kazmierczak A (2013) The contribution of local parks to neighbourhood social ties. Landsc Urban Plan 109(1):31–44. https://doi.org/10.1016/j.landurbplan.2012.05.007

Kellert S (1997) Kinship to mastery, biophilia in human evolution and development. Island Press, Washington, D.C.

Kellert SR (2002) Experiencing nature: affective, cognitive, and evaluative development in children. Children and nature. Psychol, Sociocult, Evol Investig:117151

Kellert SR (2008) A biocultural basis for an ethic toward the natural environment. The foundations of environmental sustainability: the coevolution of science and policy, vol 21. Oxford University Press, Oxford, UK, p 321-332. https://doi.org/10.1093/acprof:oso/9780195309454.003

Klotz AC, Bolino MC (2021) Bringing the great outdoors into the workplace: the energizing effect of biophilic work design. Acad Manag Rev 46(2):231–251

Konijnendijk, C C, Annerstedt, M, Nielsen, A B, & Maruthaveeran, S (2013). Benefits of urban parks. A systematic review. A Report for IFPRA, Copenhagen & Alnarp, 1–70

Kruiz H, van der Vliet N, Staatsen B, Bell R, Chiabai A, Muiños G, Stegeman I (2019) Urban green space: creating a triple win for environmental sustainability, health, and health equity through behavior change. Int J Environ Res Public Health 16(22):4403

Kuo FE, Sullivan WC, Coley RL, Brunson L (1998) Fertile ground for community: inner-city neighborhood common spaces. Am J Community Psychol 26(6):823–851

Lee ACK, Jordan HC, Horsley J (2015) Value of urban green spaces in promoting healthy living and wellbeing: prospects for planning. Risk Manag Healthc Policy:131–137

Lerner, J. (2003). Acupuntura Urbana, Rio de Janeiro-Sao Paulo0

Maas J, Van Dillen SM, Verheij RA, Groenewegen PP (2009) Social contacts as a possible mechanism behind the relation between green space and health. Health Place 15(2):586–595

Mahdiar Z, Dali MMJG (2016) Adolescent use of urban parks and their social environment consequences 12(11)

Matsuoka RH, Kaplan R (2008) People needs in the urban landscape: analysis of landscape and urban planning contributions. Landsc Urban Plan 84(1):7–19. https://doi.org/10.1016/j.landurbplan.2007.09.009

McCormack GR, Rock M, Toohey AM, Hignell D (2010) Characteristics of urban parks associated with park use and physical activity: a review of qualitative research. Health Place 16(4):712–726

McNeill LH, Kreuter MW, Subramanian SV (2006) Social environment and physical activity: a review of concepts and evidence. Soc Sci & Med 63(4):1011–1022

Michler LM, Treydte AC, Hayat H, Lemke S (2019) Marginalised herders: social dynamics and natural resource use in the fragile environment of the Richtersveld National Park, South Africa. Environ Dev 29:29–43

Mohammadi Tahroodi F, Ujang N (2022) Engaging in social interaction: relationships between the accessibility of path structure and intensity of passive social interaction in urban parks. Archnet-IJAR. Int J Arch Res 16(1):112–133

Ojo, D B (2024). Urban Design and Planning: Integrating Nature and the Built Environment

Pescosolido BA, McLeod JD, Avison WR (2007) Through the looking glass: the fortunes of the sociology of mental health. In: Mental health, social mirror. Springer, pp 3–32

Peters K, Elands B, Buijs A (2010) Social interactions in urban parks: stimulating social cohesion? Urban For & Urban Green 9(2):93–100

Priyadarsini R (2009) Urban heat Island and its impact on building energy consumption. Adv Build Energy Res 3(1):261–270
Ryan CO, Browning WD, Clancy JO, Andrews SL, Kallianpurkar NB (2014) Biophilic design patterns: emerging nature-based parameters for health and Well-being in the built environment. ArchNet-IJAR. Int J Arch Res 8(2):62
Seeland K, Dubendorfer S, Hansmann R (2009) Making friends in Zurich's urban forests and parks: the role of public green space for social inclusion of youths from different cultures. Forest Policy Econ 11(1):10–17. https://doi.org/10.1016/j.forpol.2008.07.005
Simões Aelbrecht P (2016) 'Fourth places': the contemporary public settings for informal social interaction among strangers. J Urban Des 21(1):124–152
Trombeta G, Cox SM (2022) The textual-visual thematic analysis: a framework to analyze the conjunction and interaction of visual and textual data. Qual Rep 27(6):1557–1574
Tuncer, S, Brown, B (2020, April). E-scooters on the ground: Lessons for redesigning urban micro-mobility. In Proceedings of the 2020 CHI conference on human factors in computing systems (pp. 1–14)
Turner-Skoff JB, Cavender N (2019) The benefits of trees for livable and sustainable communities. Plants, People, Planet 1(4):323–335
Ulrich RS (1993) Biophilia, biophobia, and natural landscapes. Biophilia Hypothesis 7:73–137
Villanueva K, Badland H, Kvalsvig A, O'Connor M, Christian H, Woolcock G, Goldfeld S (2016) Can the Neighborhood built environment make a difference in children's development? Building the research agenda to create evidence for PlaceBased children's policy. Acad Pediatr 16(1):10–19. https://doi.org/10.1016/j.acap.2015.09.006
Wilson OE (1984) Biophilia. Harvard University Press, Cambridge, MA
Wilson R (1994) Environmental education at the early childhood level. North American Association for Environmental Education, Washington, DC
Wolf I'D, Stricker HK, Hagenloh G (2015) Outcome-focused national park experience management: transforming participants, promoting social Well-being, and fostering place attachment. J Sustain Tour 23(3):358–381
Woolley H (2008) Watch this space! Designing for children's play in public open spaces. Geogr Compass 2(2):495–512
Woolley H, Johns R (2001) Skateboarding: the city as a playground. J Urban Des 6(2):211–230
Zhong W, Schröder T, Bekkering J (2022) Biophilic design in architecture and its contributions to health, Well-being, and sustainability: a critical review. Front Arch Res 11(1):114–141
Zhou Y, Wang M, Lin S, Qian C (2022) Relationship between children's independent activities and the built environment of outdoor activity space in residential Neighborhoods: a case study of Nanjing. Int J Environ Res Public Health 19(16):9860
Ziaesaeidi P (2025) 'Relax underneath the trees and read a book': exploring social sustainability in parks through the lens of young people. Sustainability 17(5):2138
Ziaesaeidi, P, Cushing, D F, Washington, T, Buys, L (2020, November). How youth are socially engaged in parks: a participatory-approach for understanding youth perceptions and use patterns. In IOP Conference Series: Earth and Environmental Science (Vol. 588, No. 5, pp. 052058). IOP Publishing
Ziaesaeidi P, Cushing DF, Washington T, Buys L (2023) "Just to make new friends and play with other children": understanding youth engagement within neighbourhood parks using a photo-choice tool. Landsc Urban Plan 235:104757

Finding Home in Urban Green Spaces: A Qualitative Study of Older Adults' Sense of Belonging and Attachment in Local Parks

Shima Taheri, Amirhosein Shabani, and Ali Cheshmehzangi

Abstract This study explores the role of local parks in fostering a sense of home for older adults, who often contend with physical, mental, financial, and social challenges. It also introduces a novel perspective by examining local parks through the sense of home lens, a dimension largely unexplored in the literature of local green spaces and gerontology. We conducted semi-structured interviews with 19 participants at Isfahan's Hasht Behesht Park, and a thematic analysis to identify the factors that influence the sense of home in green spaces. The findings identify emotional and social experiences, sense of place, and continuity of experience as central themes informing the sense of home in local parks. For example, emotional attachment and nostalgia reflect how parks evoke memories and comfort, linking past and present in meaningful ways. Social support and interaction reveal the importance of parks as spaces where older adults maintain relationships and build a sense of belonging. Additionally, sensory experiences—such as sounds, smells, and natural esthetics—reinforce emotional bonds with place.

Together, these elements illustrate how local parks become deeply personalized environments that support well-being and identity in later life. Although the COVID-19 pandemic-imposed limitations on participant perceptions, psychology, behavior, and social interaction—for example, restricting perceptions of safety and accessibility of public spaces, influencing psychological well-being through heightened feelings of isolation, altering behavior due to health-related cautions, and constraining opportunities for in-person social interactions—it also highlights how local green

S. Taheri
Department of Architecture, Isfahan (Khorasgan) Branch, Islamic Azad University, Isfahan, Iran

A. Shabani
Department of Urban Planning, Na.C, Islamic Azad University, Najafabad, Iran

A. Shabani (✉)
Advancement in Architecture and Urban Planning Research Center, Na.C, Islamic Azad University, Najafabad, Iran
e-mail: ashabani@phu.iaun.ac.ir

A. Cheshmehzangi
School of Architecture, Design and Planning, The University of Queensland, Brisbane, Australia

A. Cheshmehzangi et al. (eds.), *Nature in Cities, Nurturing Cities*, Urban Sustainability,
https://doi.org/10.1007/978-981-95-5487-4_11

spaces can foster social resilience. It provides insights into park design and management as personalized, sustainable, and inclusive environments that can enhance well-being, community cohesion, and nature-based solutions for healthy aging.

Keywords Sense of home · Local parks · Older adults · Sense of Place · Social support

1 Introduction

People might think older adults spend most of their time in their homes, often alone or in their children's homes or nursing homes away from the family. However, older people have discovered a new home in the embrace of trees! While studies have explored the use of parks by older adults (Ali et al., 2022; Macintyre et al., 2019; Taheri et al., 2025a; Taheri et al., 2025b; Taheri et al., 2021a, 2021b, 2023; Verderber et al., 2023; Yang et al., 2023; Xu et al., 2022), there remains a gap in understanding the unique advantages urban greenery offers them and the concept of parks as places where one can "feel at home" or consider them a "second home" has not been thoroughly explored. Furthermore, the literature on the concept of home has undergone many changes and challenges over more than 50 years, and recently, the "concept of home" and creating a "feeling at home" for special groups such as immigrants (Bunn et al., 2023; Feng and Zhu, 2022; Prato et al., 2021; Tang et al., 2022), and older adults (Verderber et al., 2023; Zhang et al., 2022) have received more attention. However, most research is limited to nursing homes (Chen et al., 2023; Eijkelenboom et al., 2017; Fleming et al., 2015; Muller et al., 2022). Home encompasses a variety of places, each associated with its own unique set of sensory experiences (Ahmet, 2013). Given the rich emotional, sensory, and social experiences that older adults gain from regular visits to neighborhood parks—such as feelings of comfort, familiarity, and belonging, these spaces can be considered a form of "second home" for many (Fig. 1). This aligns with recent studies emphasizing the role of meaningful public spaces in fostering place dependence and identity among older populations (Taheri et al., 2025b).

The urban aging population is confronting significant challenges, intensified by the COVID-19 pandemic (Buffel et al., 2023), which has highlighted issues like injustices (Crisp and Waite, 2023), small home size challenges (Hubbard, 2025), social isolation and mobility constraints (Sharifi and Khavarian-Garmsir, 2020). Social health, particularly in cities, is crucial due to the risks of loneliness and the importance of social ties (Pratt et al., 2023). The environmental and societal context (Shishegar and Boubekri, 2022) plays a pivotal role in shaping these experiences during old age. Addressing these complex challenges is essential for the health and well-being of older adults and society at large.

The concept of "home" has been extensively studied across disciplines, yet it remains a subject of debate, described as an "essentially contested concept" (Meers, 2023). The study of "home" is inherently multidisciplinary, reflecting its complexity

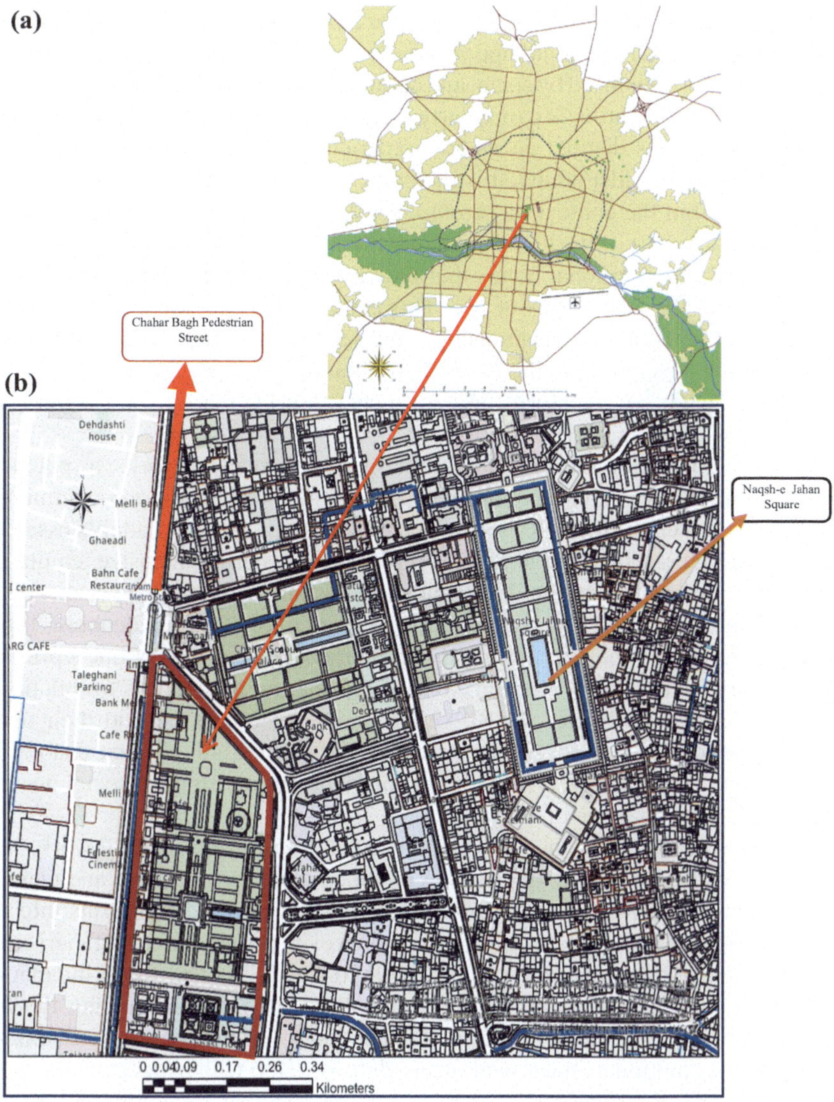

Fig. 1 **a**. The case study, Hasht Behesht park in Isfahan (Google Maps, revised by the authors, 2025) **b**. Site Plan of Hasht Behesht park and Adjacent Urban District (revised by Authors, 2025)

and the many facets it encompasses—identity, belonging, memory, and attachment (Bunn et al., 2023). Marcus's influential 1974 study posited that home is more than a physical space; it's a symbolically rich and emotionally significant place offering safety, comfort (Liu et al., 2017) and familiarity. Home's multifaceted nature is evident in its cultural representation of societal values (Meagher and Cheadle, 2020), its psychological role in shaping personal identity and emotional security (Maersk et al., 2018), and its philosophical significance as a space of continuity and authenticity (Heidegger, 2001). In contemporary views, home is increasingly recognized as a dynamic concept, influenced by social and cultural contexts (Gillon and Gibbs, 2019). This evolving interpretation is particularly meaningful for older adults, whose sense of home incorporates distinct needs and experiences. To understand the evolving concept of home for older adults, it is imperative to explore the multifaceted nature of this term within the context of gerontology. Literature in this field often underscores the significance of the home as a cornerstone for healthy aging, reflecting the deep cognitive and emotional bonds that individuals forge with their living spaces as they navigate the later stages of life. This connection is not merely physical but encompasses the memories, relationships, and sense of security (Kimic and Polko, 2022) that collectively define the essence of "home." As such, the home becomes more than a shelter—it transforms into a sanctuary of familiarity and comfort that supports the well-being of older adults.

Extending this notion, introducing public spaces, particularly parks, as potential "second homes" offers a compelling perspective on socialization and community engagement (Cui et al., 2024; Rogers et al., 2020) among the older adult's essential social infrastructure (Taheri et al., 2025b), pivotal in fostering inclusive, equitable, and vibrant (Layton and Latham, 2022) and more sustainable (Campbell et al., 2022). Drawing from Ray Oldenburg's influential concept of third places (Oldenburg, 1989) which are vital public spots for informal social interactions outside of one's primary and secondary environments, we can envision parks as arena catalysts for relational curiosity (Phillips and Evans, 2018). Urban parks can evolve beyond their usual function as places of leisure, becoming essential "second homes" that enable the older people to engage actively with their community, establish fresh social ties and flourish within the embrace of nature.

To effectively highlight the gaps and articulate the novelty of this work, it is essential to synthesize the existing studies on older adults' use of parks and identify where further research is needed. Previous studies have provided insights into the benefits of park usage for older adults, such as improved physical health (Taheri et al., 2025a), mental well-being (Taheri et al., 2019, 2021a, 2023), and social interaction (Taheri et al., 2025b). However, there are notable gaps in understanding the full spectrum of older adults' experiences in these settings. For instance, there is limited research on the emotional attachment older adults develop with parks and how these spaces can serve as extensions of their home environment, providing a sense of belonging and community.

The novelty of the current work lies in its holistic approach to understanding the concept of parks as "second homes" for older adults. This study aims to bridge the identified gaps by synthesizing existing literature with new empirical data and examining the multifaceted relationship between older adults and park usage in the post-pandemic era. It seeks to provide a deeper understanding of how parks can be designed and managed to meet the specific needs of older adults, fostering environments that are not only safe and accessible but also nurturing and conducive to active, social lifestyles. This work contributes a fresh perspective to the gerontological discourse on "home" and public spaces, offering actionable insights for urban planners and policymakers to create age-friendly cities.

2 Research Methodology

This study employed a qualitative research design to explore older adults' sense of home and belonging in local parks. Data was collected through semi-structured interviews with 19 participants in Hasht Behesht Park, Isfahan. The interviews provided rich, in-depth accounts of participants' experiences, perceptions, and attachments to the park. For data analysis, we applied Braun and Clarke's (2006) thematic analysis method, which involves systematically identifying, analyzing, and reporting patterns (themes) within qualitative data. This process included familiarization with the data, generating initial codes, searching for and reviewing themes, and finally defining and naming them to interpret the underlying meanings.

2.1 The Case Study: Hasht Behesht Park in Isfahan, Iran

Isfahan is Iran's third most populous city. The city's cultural heritage is integral to its World Heritage Site (WHS) identity (Hazegh et al., 2025; Taheri et al., 2024), which plays a key role in fostering a sense of place and belonging (Salehi et al., 2023). Historical gardens located along the Chahar Bagh axis are significant natural and historical features within the city's landscape. Among them is the Hasht Behesht Garden, now recognized as Hasht Behesht Park, a notable example of traditional Iranian gardens. Due to its frequent use by older people, this park attracted the authors' attention for further investigation as a case study. Originally this urban green space designed during the seventeenth century as part of the royal complex surrounding the Hasht Behesht Palace, the park retains a formal Persian garden layout, including symmetrical pathways, water features, and mature trees that offer shaded, contemplative spaces.

Hasht Behesht Park is located near Chahar Bagh Street, one of the most historically significant and well-known thoroughfares in both Isfahan and Iran. Dating back to the Safavid era, Chahar Bagh served as the principal north–south axis of the city, forming the spine of a grand urban plan that linked a series of royal gardens and palaces. As the hinge of Safavid landscape and urban design, this street played a central role in shaping the spatial and cultural identity of Isfahan. Over time, Chahar Bagh and its surrounding gardens have undergone substantial physical transformations, with many of the original garden structures lost or altered. Today, Chahar Bagh has been converted into a pedestrian walkway, and Hasht Behesht Park stands as one of the few remaining elements of the Safavid garden network. This historical continuity, along with its central location and preserved features, positions the park as a meaningful and accessible green space that continues to attract daily visitors, especially older adults.

2.2 Data Collection

Data was collected through semi-structured interviews with 19 older adults who frequently visited the park. The interview process was meticulously organized into four distinct sections and lasted between 40 to 60 minutes.

The first section focused on introductory questions, such as personal characteristics, attendance and accompaniment. This section aimed to establish a comfortable atmosphere and allow participants to ease into the conversation. The second section shifted subtly toward more reflective questions, inviting participants to disclose their motivations for visiting the park. Subsequently, the third section delved deeper to align with the specific objectives of the study—participants were asked to contemplate why they favored park visits overstaying at home and "Is there anything in the park that makes you want to come so much because it attracts your attention?" These questions allowed participants to reflect on their park experiences and share their thoughts and feelings more detailed and nuancedly. Finally, the fourth section of the interview was a platform for participants to express unspoken issues of their choice. They were encouraged to share personal stories, describe the park, or relate any other experiences they wanted to discuss.

2.3 Participants

The participants in our study were selected from older adults who regularly visited Hasht Behesht Park. We used a combination of criteria-based and snowball sampling to identify older adults who met specific inclusion conditions. Participants had to be over 65 years of age and demonstrate a sequence of experience, meaning they had a history of repeated visits to the park over an extended period (at least one year). "Sufficient time in the park" referred to individuals who spent a minimum of 30 minutes in the park during each visit and visited the park at least three times per week. Additional criteria included a willingness to participate in a face-to-face, in-depth interview and the ability to comply with public health protocols (e.g., mask-wearing, social distancing) during the research process. In total, 19 interviews were conducted, after which data saturation was reached, indicating that no new significant themes were emerging from the data and our research questions were adequately addressed.

2.4 Data Analysis

All interview transcripts were systematically imported and managed in MAXQDA Analytics Pro (version 20.4.1) by the first author, ensuring a structured and rigorous foundation for the subsequent coding and analysis process. For each interview, the authors systematically and sequentially organized the codes inductively into sub-themes and four main themes. The first author piloted the coding and classification processes and discussed the applied procedures with the second and third author. Based on our research question: How and why older people develop a sense of home in urban park settings? themes were defined. All authors discussed and agreed on the main themes and sub-themes.

3 Results

Our analysis revealed four interconnected yet analytically distinct themes: social experience, emotional experience, sense of place, and continuity of experience. While these themes are interrelated and may occasionally intersect, each represents a unique dimension of older adults' interactions with Hasht Behesht Park. This thematic framework is intentionally flexible to capture the nuanced, layered, and evolving nature of the participants lived experiences.

3.1 Social Experiences

Hasht Behesht Park facilitates a wide range of social experiences—such as social interaction, support, vitality, and the formation of strong social networks—through both its physical layout and sociocultural atmosphere. The park's open and accessible design, with ample shaded seating areas, interconnected walking paths, water features, and proximity to the pedestrianized Chahar Bagh Street, encourages spontaneous gatherings and prolonged stays. These spatial qualities, combined with the park's historical and central location, attract diverse user groups and provide a familiar and comfortable environment for older adults to engage in meaningful social activities.

Older participants reported engaging in various shared practices, including conversation, communal eating, storytelling, recreation, walking together, and even informal teaching and learning. These everyday interactions take place in specific zones of the park that have become informal "hangouts" stable gathering points shaped by routine presence, mutual recognition, and long-standing social bonds. As one male participant in his late 60 s explained: *"Now Mr.... does not come, we ask each other again, he was sick, we all got upset and went to visit him together, we look out for each other"; "Mr. we interact with them in the park as a spiritual father"*

Such experiences are facilitated not just by the presence of peers but also by the park's calming atmosphere, historical significance, and walkable structure, which together foster a strong sense of familiarity and emotional safety. These factors mirror aspects of a home environment, where mutual responsibility, empathy, and emotional care are expected. For some older adults, the park becomes a "second home," a space where social connections extend beyond casual interaction into meaningful social support networks.

Some older participants referred to Hasht Behesht Park as a "second home," highlighting its emotional significance and the comfort it provides. This description takes on deeper meaning when contrasted with the challenges many of them face in their actual homes. For instance, several participants reported living alone or experiencing emotional neglect within their families, making opportunities for meaningful interaction at home scarce or absent. As one 68-year-old woman put it: "What should I do when I go home? I'm miserable—I just want to stay out for two more hours." Another 70-year-old man shared: "I am alone at home; no one talks to me! That's why I come to the park—to be at ease, both for myself and for my family."

Interestingly, even participants who did not seek out frequent conversation valued the park for its natural environment. The presence of trees, open space, and tranquility allowed for a kind of restorative solitude that supported emotional resilience and psychological well-being, something they felt was lacking in their home setting. As a retired 72-year-old man noted: "Because of these trees, I come to the park without wanting to have friends." These reflections reveal that the park offers not only social engagement but also vital emotional refugees, especially for those who lack support or connection in their private domestic spaces. These narratives underscore how the park functions as a socially enabling environment, providing both companionship and psychological support. It fosters a context where older individuals not only engage in direct social interaction but also feel recognized and valued within informal networks. Even those who preferred solitude found a form of social presence in the shared use of space and the soothing influence of nature.

Several participants in this study described feeling overwhelmed or emotionally burdened by everyday family dynamics or the limitations of home life, leading them to seek relief in the park environment. As a 67-year-old woman explained: "*When my son angers me, I come here alone to calm down.*" Similarly, a 74-year-old man noted: "*For me, coming to the park is to spend time and get away from home.*" These statements suggest that Hasht Behesht Park offers an alternative space—outside the confines of domestic life—where older adults can find emotional relief and a change of atmosphere. Although some individuals initially came to the park alone, many found comfort in informal social groups or "hangouts" that emerged organically in specific areas of the park. These gatherings, often rooted in shared routines and life experiences, allowed participants to engage in unstructured, spontaneous conversations. Topics ranged from everyday events to family matters, creating a low-pressure setting for emotional expression and interpersonal connection. As one 71-year-old retired teacher described: "*Here, retirees relieve themselves from the troubles of their lives!*" The park's physical environment, including accessible paths, shaded seating, and calm, natural settings supported these social interactions while also encouraging light physical activity, such as walking. These shared activities not only fostered a sense of connection but also contributed to a broader feeling of belonging. For many, the park evoked a "second home," not in a literal sense, but as a space that offered familiarity, routine, and social warmth absent in other areas of daily life.

These findings reinforce the idea that urban green spaces can serve as supportive and restorative environments for older adults. Through a combination of physical design, natural elements, and social affordances, parks like Hasht Behesht provide critical opportunities for social engagement, emotional well-being, and community-building in later life.

3.2 Emotional Experience

Participants described a variety of emotional experiences associated with their time in Hasht Behesht Park. These included feelings of peace, inspiration, safety, and comfort. The park's natural and esthetic qualities—such as its mature trees, seasonal change, flowing water, and birdsong—were frequently mentioned as sources of emotional relief and mental clarity. These multisensory stimuli played a key role in shaping how older adults connected with space. As 76-year-old man shared: "*Here we feel the seasons and time passing, and I got inspiration for my poetry book.*" According to Weinberger et al. (2022), esthetic responses to natural environments are influenced by psychological factors such as fascination, coherence, and hominess. These dimensions were reflected in participants' reflections, especially in how the park provided a calming alternative to emotionally restrictive home environments. Emotional comfort in the park was not always dependent on social interaction—some participants valued the opportunity for solitary reflection and the freedom to engage on their own terms.

A 70-year-old woman remarked: "*I feel unfortunate when I want to go home from here, but I hope that I will be back soon.*" However, participants also identified disturbing and disruptive elements that negatively impacted their emotional experience. For instance, noise pollution caused by passing motorcycles, construction noise from nearby development projects, and the occasional presence of individuals struggling with addiction, particularly in specific, less-monitored areas of the park—were frequently cited as sources of discomfort or anxiety. These interruptions detracted from the otherwise calming atmosphere, limiting the park's potential as a fully restorative space. As one participant noted: "*Sometimes the noise from outside just ruins the peace—I come here for quiet, but it's not always quiet.*"

3.3 Sense of Place

In this study, sense of place was assessed through emotions, behavioral exclusivity, and cognitive forms of attachment (Lin and Lockwood, 2014; Salehi et al.,

2023). These factors contribute to developing older people's "feeling at home" as a multidimensional and complex phenomenon:

- Cognitive component: Familiarity with the park's historical, cultural, environmental, social, and natural structures and place image help develop a sense of place. For many seniors, this knowledge reaches several decades, and they have a clear mental picture of the park's changes during this period. This familiarity has enhanced comfort, created an intimate atmosphere, and depended on their connection with the park.*"I have been in Isfahan for 50 years and come to this park!"*
- Spatial component: This includes the park's historical and cultural identity, attractiveness and natural beauty. Trees, flowers, and animals create a calm, safe, and renewed atmosphere for older adults to gather and communicate in the park.*" I love this park so much!"*
- Emotional component: Many positive emotional, social, and sensory behaviors have fostered a positive emotional connection. Participating in activities and events, expressing emotions, enjoying the view, and establishing pleasant communication with others serves as the basis for forming memories and the sense of dependence on natural-human elements, and they encourage older people to return to the park. For many older adults, these positive experiences even go back to childhood and adolescence, which has caused a deep emotional and psychological attachment to the park; even after they have left the park, they feel a strong emotional connection with the park of their childhood. It shows the importance of cross-generational transmission, place identity and attachment (Frost and Catney, 2020; Taheri et al., 2025b).*"We are attached to this park, we love it... yes, wherever we are with our friends, we have fun staying there." Lost"; "I will come every day even if there is an earthquake."*

Consequently, sense of place is a multidimensional concept that plays a vital role in shaping seniors' relationship with the park and the "sense of being at home." The cognitive, emotional, and behavioral components of the sense of place are closely intertwined and shaped by various factors, including personal experiences of individuals, cultural background, social interactions, and environmental context (Fig. 2). Since one of the essential characteristics for calling a place "home" is the presence of a sense of place, dependence, and belonging to that place, by proving the high level of sense of place in Hasht Behesht Park, the "sense of being at home" of older people can be better understood.

3.4 Long-Term Use, Shared Memory, and Place Attachment

The deepening of relationship with the place and with seniors in Hasht Behesht Park reflects Relph's (2008) concept of three levels of behavioral innerness: behavioral innerness (attention to and understanding of older people's familiarity with the park), emotional innerness (efforts toward empathy, protection, place meanings,

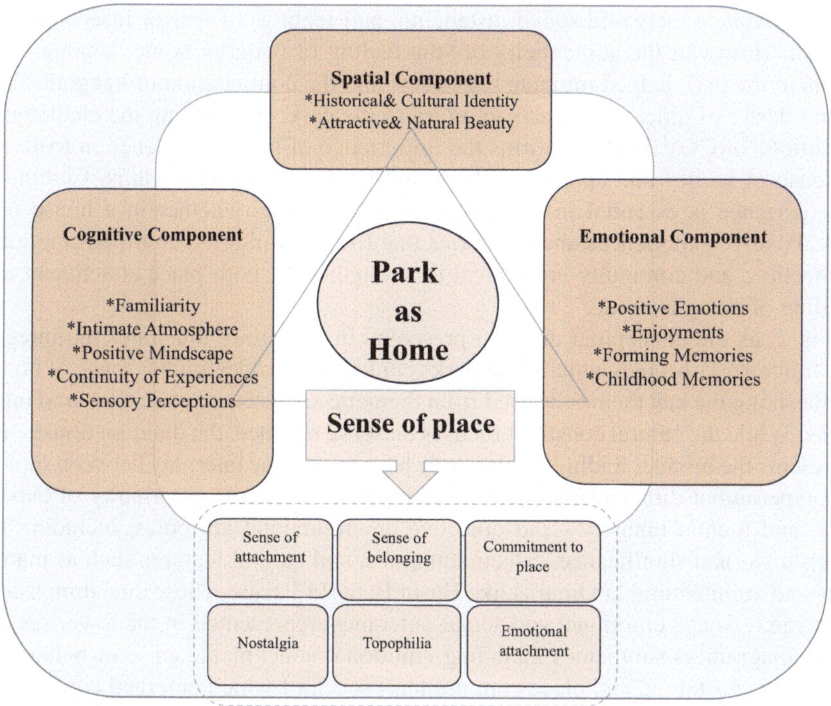

Fig. 2 Sense of place and creating a sense of home in the park

and identification with the events of the place), and existential innerness (a sense of understanding the meaning of the place and mental belonging). Previous positive experience, along with the continuity of older adults' presence in friendship groups and the formation of shared memories provide a favorable foundation for long-term engagement and define the park as a setting for the continuity of memory. Seniors recalled the importance of parks during special periods, such as engagements, school years, and family gatherings. *"We used to date here with my fiancé"*, *"The memories we had of this park since childhood, especially with my grandfather..."* Spatial experiences for older adults, whether in exploring a place (such as finding a historical tree), writing poetry, singing in a group, or playing music in a particular area of the park over a long period, became personalized and shared in various ways. Notably, majority of older people reported also visiting other parks concurrently with Hasht Behesht Park.

The continuity and sequence of older adults' experience often reflect long-term experience-ranging from one year to several decades— and most participants reported regular patterns in both the days and hours of their visits. This continuity has played a significant role in deepening their sense of belonging and attachment to the place. *"We are attached to this park. We have been coming here for 40 and 50 years with our friends..."* The impact of the recent pandemic-such as reduced park attendance increased social distancing, and feelings of fear or insecurity-was evident. However, the authenticity of "the feeling of being at home "among older adults in the park helped mitigate the effects and the continuation of hangouts. The strong desire of older individuals to return to the park, even during the challenging conditions of COVID-19, confirms the importance of lived experience in fostering a "sense of home" and enhancing the social resilience of older adults. Continuity of experience is essential in creating a sense of home- whether in a house or a park. Positive experiences and memories that foster familiarity, comfort, emotional connection, and continuity are critical to strengthening both place attachment and "feeling of being at home."

Fig 2. as the conceptual diagram presented in this study illustrates the integrative framework through which local parks contribute to older adults' sense of home, synthesizing the key themes derived from thematic analysis of semi-structured interviews. While the central construct focuses on sense of place, the diagram holistically represents the broader findings of the study by mapping the interplay between subjective experiential dimensions—such as memories, familiarity, continuity of experience, and mental imagery—and objective environmental attributes, including the park's historical significance, esthetic qualities, and natural features such as mature trees and architectural landmarks like Hasht Behesht Palace. These dual dimensions converge to shape emotional and social outcomes, represented in the lower section of the diagram as subthemes including emotional attachment, sense of belonging, topophilia, nostalgia, and place commitment—each of which emerged consistently in participants' narratives. The organization of the diagram reflects the qualitative data structure, demonstrating how personal, sensory, and social experiences interact within a meaningful physical environment to foster a deep, affective connection to place. Far from being limited to a single theme, the visual model encapsulates the complexity of the lived experience in urban green spaces, aligning directly with the thematic categories discussed in the results and discussion sections. This integrative representation aims to advance understanding of how parks function not merely as recreational spaces, but as extensions of home and identity in later life.

4 Discussion

Older adults' presence at Hasht Behesht Park is significantly influenced by their sense of belonging and interactions with friends. The sense of identity and attachment to this park—as components of the sense of place—along with positive feelings, memories, and experiences, serve as important foundations for initiating the experience of "feeling at home." Place identity functions as both an identification tool and a defensive mechanism, helping to maintain continuity in the life of older adults (Li and Zhang, 2021). A stronger sense of place identity is also associated with more resilient communities (Sobhaninia et al., 2023).

As described in Rafiei and Gifford's, (2023), creating comfort and a sense of belonging provides the context for more positive experiences. In this park, the sense of place—relational in nature-has led older adults to internalize the landscape, transforming it forms a routine public space into a meaningful place with the qualities of a *home*. The underlying reason for this feeling and the continuity of presence can be found in their social experiences, particularly through older adults' narratives of "the feeling of being at home." Many participants linked their ongoing presence to meaningful social interactions and supportive and empathetic social networks.

Within the familiar pattern of "feeling at home," the theme of sensory-emotional experiences emerges as a catalytic force for other themes. As older adults' satisfaction with the park increased- along with feeling of autonomy and safety, the sensory-emotional aspects became a driving engine in shaping this connection. The continuation of "feeling at home" experience, rooted in a stable and deep connection between self and park, has been reinforced through previous positive experiences and consistent use.

This emotional and social connection suggests that the experience of "feeling at home" among older adults is shaped by both relational and environmental dimensions and is not necessarily exclusive to Hasht Behesht Park. Since many seniors also frequent nearby parks simultaneously, it is likely that similar experiences could be cultivated in other local green spaces that support these emotional and social dynamics.

As people age, the home and feeling "being at home" become more increasingly significant (Amián et al., 2021; Blackler et al., 2023). The relationship between the home and the individual is complex, and the home's perceived social, emotional, and cognitive dimensions directly affect older adults' health and well-being (Eriksson et al., 2022). Home is not only a physical space but also an experiential one. Through the experiences of living in different places, people develop a sense of what makes a place feel like home.

For older adults, the experience of "home" and "homeness" is multidimensional. It includes sensory, emotional, and social elements, and their continuity fosters a deep connection—whether to a traditional residence or to a familiar public space such as a local park. Affective experience refers to the feelings of empathy, sympathy, support, comfort, security, and attachment that individuals associate with their homes. Social experience relates to the interpersonal relationships and social networks people build in what may be considered a "home-park." The cultural dimension of the home involves the meanings people attribute to their lived environments, as well as the beliefs and traditions inherited from their communities. These cultural layers can significantly shape how the concept of "home" is perceived and experienced.

This understanding of the concept of home—grounded in the capacity for interaction, social participation (Amián et al., 2021), sense of place, social equity (Larimian and Sadeghi, 2021) can inform expectations of a sustainable design approach to the home. Both the home and the local park serve as vital resources for older adults' well-being, identity, and social connection. Together, they become integral components of a sustainable senior care strategy, promoting physical and mental well-being and providing opportunities for connection and socialization.

Several studies have confirmed that parks function as essential health supporting environments (Taheri et al., 2021b; Zhang et al., 2015), especially when understood through the lens of "home" as a supportive space. It is well established that natural environments possess rich sensory and psychological features that make them particularly conducive to restorative experiences (Macaulay et al., 2022; Moll et al., 2022; Rapuano et al., 2022; Taheri et al., 2019). These qualities are especially valuable in reducing stress and anxiety—conditions that became increasingly prominent during the COVID-19 pandemic and prompted cities to prioritize urban resilience in their planning efforts (Amirzadeh et al., 2023).

Studies examining the relationship between the psychological dimension of home and residential environments for older adults confirm—and in some cases recommend—that contact with nature, particularly through access to local green spaces, can have a significant impact. From a psychological perspective, the feeling of being at home within one's living environment (Rosenberg et al., 2021) provides a secure foundation for the formation of identity and pursuit of self-actualization (Seo and Mazumdar, 2011).

Looking more deeply, the meaning of home is not inherent in objects; rather, it is produced through continuous, everyday interactions between the user and the space (Falk et al., 2013). The feeling of being at home emerges through this interactive process—from objectivity to subjectivity—and is shaped by the individual's age, social context, and cultural background. This makes the concept of home inherently complex and fluid.

In some cases, such as in Lovatt (2018) study, the nursing home has replaced the traditional home, demonstrating how spatial meaning can shift based on user interaction and relational dynamics. In terms of scale, "home" can refer to a wide spectrum—from one's childhood house to a city, a homeland, or even the entire planet. Thus, the concept of home and the feeling of being at home is both comprehensive and multifaceted.

In recent years, the literature on home has focused on two main groups: first, immigrants, in relation to the loss of their original home and their efforts to create or adapt to a new one; and second, older adults living in care centers, who face similar challenges in adapting to a new phase of life. For older adults, being at home—particularly toward the end of life—holds deep personal significance (Jellard and Bell, 2021), making the experience of changing or losing one's home especially difficult.

In Z. Zhang et al., (2022), the challenge of residence and the sense of home was explored for older adults' immigrants. Similarly, as noted by Nieboer and Cramm (2024), such challenges are often more impactful for older adults than for other groups. Therefore, reinterpreting urban green spaces as a "second home" for older immigrants can help reduce social tensions and facilitate greater adaptation to host communities.

For some older adults, choosing or being compelled to live in nursing homes or care centers marks a significant transition. In these settings, efforts are often made to evoke and support the experience of "being at home" (Klaassens and Meijering, 2015; Muller et al., 2022; Seo and Mazumdar, 2011). Incorporating natural elements into such environments has been shown to enhance this feeling (Eijkelenboom et al., 2017). However, creating a new sense of home in an unfamiliar place is inherently complex and multi-layered (Seo and Mazumdar, 2011). Without a doubt, supportive factors—such as access to nature and the integration of green spaces—can significantly ease this process and promote smoother emotional and social adjustment.

Although Jellard & Bell's study (2021) focuses on the beach as a home-like environment, it also supports the possibility of experiencing a similar feeling in urban green spaces. Just as users can enhance beaches as spaces of public ownership without facing legal obstacles (McArdle and Byrne, 2022), urban green spaces can serve similar role for older adults. In times of crisis—such as COVID-19 pandemic—instead of restricting and blocking older adults' access to local parks, we align with the recommendations of (Tabrizi et al., 2023), in proposing preventive solutions and prioritizing safe, continued access for older users. The loss of a home-like environment during such periods can cause irreparable challenges for older adults (Jellard and Bell, 2021).

One promising solution involves leveraging emerging technologies, such as capacity management and control applications within the framework of a "smart environmentality" (Yu et al., 2024) which can help reduce inequitable access to green spaces during pandemics. Access to nature and sensory stimulation is a key component in fostering a sense of home for older adults, particularly those with Alzheimer's disease (Fleming et al., 2015).

The flexible and inclusive nature of parks not only fosters strong social relationships (Taheri et al., 2025b)—one of the most complex contributors to a sense of home (Chen et al., 2023)—but also provides space for routine, everyday activities (Falk et al., 2013), which are especially beneficial in supporting older adults with cognitive challenges. Factors that enhance the feeling of being at home, such as familiarity, sensory comfort, and social engagement, also contribute to preserving identity and promoting well-being (Eijkelenboom et al., 2017; Tucker, 2010). Therefore, local green spaces can create a new perspective for the design and planning of age-friendly spaces and mitigate the challenges of the aging period linked to the concept of home.

This research explores how local parks can play a complementary role to the concept of "home" in addressing aging-related challenges. According to our findings, older adults use local parks primarily to meet their specific needs, challenges, and preferences, further reinforcing the priorities outlined in age-friendly city research (Buffel et al., 2023). Planners, designers, and landscape architects are thus encouraged to maximize parks' potential as platforms for fostering social experiences, sense of place, sensory-emotional connections, and memory formation, while supporting regular presence so that parks begin to feel like home.

Just as people feel attachment to their homes because they offer shelter, security, and stability, particularly during times of crisis, such as COVID-19 (Meagher and Cheadle, 2020)—local parks, when envisioned as "second homes," can similarly promote older adults' well-being. The behavioral, social, and meaningful dimensions of home have been closely linked to older adults' mental and physical health (Kylén et al., 2017); identifying these aspects in community parks as complementary roles to home can be equally beneficial to older adults.

Our research also reveals that childhood experiences and parental attitudes toward home deeply shape older adults' current feelings of homeness. Considering the intergenerational impacts of nature experiences (Taheri et al., 2025b; Vitale et al., 2022; Wu et al., 2023) as well as nature's role in strengthening children's social skills and well-being (Meidenbauer et al., 2019), and the importance of place in children's everyday lives (Moskal, 2015), we suggest that efforts to create a "feeling of being at home" in parks should also include children—as both the current young generation and future older adults.

There is a clear and robust link between older adults and their sense of home (Zhang et al., 2022), and if the environment is friendly to older adults, it is likely to be close to everyone (Torku et al., 2022). However, people react to environments very differently (Weinberger et al., 2022) and therefore, the same research can be conducted on other types of urban green spaces to learn the benefits of "being at home" for different users. The concept of "being at home" can be explored by observing older people regularly in local parks and its role in strengthening neighborhood social resilience.

5 Conclusions

This study, conducted in Hasht Behesht Park, underscores the critical role that local urban parks can play in supporting the health and well-being of older adults. By exploring the nuanced and multi-dimensional relationship between older individuals and public green spaces, the research demonstrates how parks can serve as meaningful extensions of the home environment. The findings highlight that the concept of "feeling at home," traditionally associated with private domestic settings—can be cultivated in well-designed, socially responsive, and emotionally resonant public spaces. The results reveal that emotional attachment, social interaction, sensory engagement, and continuity of experience all contribute to fostering a strong sense of place and belonging among older adults.

This sense of "being at home" in the park supports not only psychological and social resilience but also broader goals of age-friendly urbanism. As cities face the dual challenges of aging populations and increasing urbanization, reimagining local parks as socially inclusive and emotionally supportive spaces becomes essential. In this context, urban planners, designers, and policymakers are urged to recognize the transformative potential of public green spaces in enhancing quality of life for older residents. Integrating principles of emotional sustainability and place-based belonging into the planning of urban green infrastructure can help cities become more equitable, inclusive, and resilient for all generations.

5.1 Policy Implications

Age-Friendly Urban Planning: Our findings underscore the importance of integrating age-friendly principles into urban planning, emphasizing the necessity for accessible and well-designed local parks. Policymakers and planners are urged to recognize parks as potential "second homes" for older adults, prioritizing features that enhance social experiences, sensory-emotional links, and a sense of place.

Technology Integration for Safety: To address the challenges posed by events like the COVID-19 pandemic, there is a pressing need to integrate technological solutions, such as capacity management and control applications. Considering the recent applications of artificial intelligence in urban planning (Shabani et al., 2025), these technological interventions not only enhance the efficiency of urban green spaces but also ensure safe access for older adults, mitigating social isolation and maintaining their sense of home.

Intergenerational Focus: Acknowledging the intergenerational impact of nature experiences, our study suggests that efforts should extend beyond the current older adult population. Urban planners and community developers are encouraged to create environments by fostering a "feeling of being at home" for children, laying groundwork for future seniors.

Declaration of Generative AI and AI-Assisted Technologies in the Writing Process While preparing this work, the authors used relevant AI tools and Grammarly to improve the quality and efficiency of their writing.

References

Ahmet A (2013) Home sites: the location(s) of 'home' for Young men. Urban Stud 50(3):621–634

Ali MJ, Rahaman M, Hossain SI (2022) Urban green spaces for elderly human health: a planning model for healthy city living. Land Use Policy 114:105970

Amián JG, Alarcón D, Fernández-Portero C et al (2021) Aging living at home: residential satisfaction among active older adults based on the perceived home model. Int J Environ Res Public Health 18(17):8959

Amirzadeh M, Sobhaninia S, Buckman ST et al (2023) Towards building resilient cities to pandemics: a review of COVID-19 literature. Sustain Cities Soc 89:104326

Blackler A, Craig C, Brophy C et al (2023) Making a "home" into a home: how design of aged-care homes impacts residents. J Aging Stud 65:101135

Braun V, Clarke V (2006) Using thematic analysis in psychology. Qual Res Psychol 3(2):77–101

Buffel T, Yarker S, Phillipson C et al (2023) Locked down by inequality: older people and the COVID-19 pandemic. Urban Stud 60(8):1465–1482

Bunn M, Samuels G, Higson-Smith C (2023) Ambiguous loss of home: Syrian refugees and the process of losing and remaking home. Wellbeing, Space Soc 4:100136

Campbell LK, Svendsen E, Johnson M et al (2022) Activating urban environments as social infrastructure through civic stewardship. Urban Geogr 43(5):713–734

Chen J, Gramegna SM, Biamonti A (2023) A sense of home for people with dementia in a long-term care facility: a design perspective. Health Place 79:102957

Crisp R, Waite D (2023) Building back better in urban contexts through a dual ethics of justice and care. Urban Geogr 44(10):2271–2279

Cui H, Maliki NZ, Wang Y (2024) The role of urban parks in promoting social interaction of older adults in China. Sustainability 16(5):2088

Eijkelenboom A, Verbeek H, Felix E et al (2017) Architectural factors influencing the sense of home in nursing homes: an operationalization for practice. Front Arch Res 6(2):111–122

Eriksson E, Wazinski K, Wanka A et al (2022) Perceived housing in relation to retirement and relocation: a qualitative interview study among older adults. Int J Environ Res Public Health 19(20):13314

Falk H, Wijk H, Persson L-O et al (2013) A sense of home in residential care. Scand J Caring Sci 27(4):999–1009

Feng D, Zhu H (2022) Migrant resettlement in rural China: homemaking and sense of belonging after domicide. J Rural Stud 93:301–308

Fleming R, Kelly F, Stillfried G (2015) 'I want to feel at home': establishing what aspects of environmental design are important to people with dementia nearing the end of life. BMC Palliat Care 14(1):26

Frost D, Catney G (2020) Belonging and the intergenerational transmission of place identity: reflections on a British inner-city neighbourhood. Urban Stud 57(14):2833–2849

Gillon C, Gibbs L (2019) Coastal homemaking: navigating housing ideals, home realities, and more-than-human processes. Environ Plan D: Soc Space 37(1):104–121

Hazegh N, Shabani A, Taheri S (2025) Culture-led regeneration based on eventfulness with an emphasis on art festivals (case study: the international film festival for children and youth in the central fabric of Isfahan city Iran). J Urban: Int Res Placemaking Urban Sustain:1–20

Heidegger (2001) Poetry, language, thought. HarperCollins

Hubbard P (2025) Small is beautiful? Making sense of 'shrinking' homes. Urban Stud 62(1):187–203

Jellard S, Bell SL (2021) A fragmented sense of home: reconfiguring therapeutic coastal encounters in Covid-19 times. Emot Space Soc 40:100818

Kimic K, Polko P (2022) The use of urban parks by older adults in the context of perceived security. Int J Environ Res Public Health 19(7):4184

Klaassens M, Meijering L (2015) Experiences of home and institution in a secured nursing home ward in The Netherlands: a participatory intervention study. J Aging Stud 34:92–102

Kylén M, Schmidt SM, Iwarsson S et al (2017) Perceived home is associated with psychological Well-being in a cohort aged 67–70 years. J Environ Psychol 51:239–247

Larimian T, Sadeghi A (2021) Measuring urban social sustainability: scale development and validation. Environ Plan B: Urban Anal City Sci 48(4):621–637

Layton J, Latham A (2022) Social infrastructure and public life–notes on Finsbury Park, London. Urban Geogr 43(5):755–776

Li X, Zhang T (2021) Place identity and older residents' coping strategies while ageing in declining neighbourhoods of urban China. J Environ Psychol 78:101692

Lin C-C, Lockwood M (2014) Forms and sources of place attachment: evidence from two protected areas. Geoforum 53:74–81

Liu Y, Dijst M, Geertman S (2017) The subjective Well-being of older adults in Shanghai: the role of residential environment and individual resources. Urban Stud 54(7):1692–1714

Lovatt M (2018) Becoming at home in residential care for older people: a material culture perspective. Sociol Health & Illn 40(2):366–378

Macaulay R, Johnson K, Lee K et al (2022) Comparing the effect of mindful and other engagement interventions in nature on attention restoration, nature connection, and mood. J Environ Psychol 81:101813

Macintyre VG, Cotterill S, Anderson J et al (2019, 1994) "I would never come here because I've got my own garden": older adults' perceptions of small urban green spaces. Int J Environ Res Public Health 16(11)

Maersk JL, Cutchin MP, la Cour K (2018) Identity and home: understanding the experience of people with advanced cancer. Health Place 51:11–18

McArdle R, Byrne M (2022) Rootlessness: how the Irish private rental sector prevents tenants feeling secure in their homes and tenant's resistance against this. Geoforum 136:211–218

Meagher BR, Cheadle AD (2020) Distant from others, but close to home: the relationship between home attachment and mental health during COVID-19. J Environ Psychol 72:101516

Meers J (2023) 'Home' as an essentially contested concept and why this matters. Hous Stud 38(4):597–614

Meidenbauer KL, Stenfors CUD, Young J et al (2019) The gradual development of the preference for natural environments. J Environ Psychol 65:101328

Moll A, Collado S, Staats H et al (2022) Restorative effects of exposure to nature on children and adolescents: a systematic review. J Environ Psychol 84:101884

Moskal M (2015) 'When I think home I think family here and there': Translocal and social ideas of home in narratives of migrant children and young people. Geoforum 58:143–152

Muller A, Missotten P, Adam S (2022) Transforming nursing home culture: opinions of older people. A cross-sectional study in Belgium. J Aging Stud 61:101020

Nieboer AP, Cramm JM (2024) Growing old outside of one's home country: Well-being needs for aging in place among Turkish people in The Netherlands. Cities 150:105065

Oldenburg R (1989) The great good place: cafés, coffee shops, community centers, beauty parlors, general stores, bars, hangouts, and how they get you through the day. Paragon House

Phillips R, Evans B (2018) Friendship, curiosity and the city: dementia friends and memory walks in Liverpool. Urban Stud 55(3):639–654

Prato L, Torregrossa M, Ramis Y et al (2021) Assembling the sense of home in emigrant elite athletes: cultural transitions, narrative and materiality. Psychol Sport Exerc 55:101959

Pratt G, Johnston C, Johnson K (2023) Robots and care of the ageing self: an emerging economy of loneliness. Environ Plan A: Econ Space 55(8):2051–2066

Rafiei S, Gifford R (2023) The meaning of the built environment: a comprehensive model based on users traversing their university campus. J Environ Psychol 87:101975

Rapuano M, Ruotolo F, Ruggiero G et al (2022) Spaces for relaxing, spaces for recharging: how parks affect people's emotions. J Environ Psychol 81:101809

Rogers WA, Ramadhani WA Harris MT (2020) Defining Aging in Place: The Intersectionality of Space, Person, and Time. Innovation in Aging Albert SM (ed.) 4(4)

Rosenberg A, Keene DE, Schlesinger P et al (2021) "I don't know what home feels like anymore": residential spaces and the absence of ontological security for people returning from incarceration. Soc Sci & Med 272:113734

Salehi M, Taheri S, Shabani A (2023) Regenerating historical districts through tactical urbanism: a case study of Sarpol neighborhood in Isfahan Province, Iran. Front Arch Res 12(6):1234–1245

Seo YK, Mazumdar S (2011) Feeling at home: Korean Americans in senior public housing. J Aging Stud 25(3):233–242

Shabani A, AlWaer H, Taheri S (2025) Artificial Intelligence Revolution in Urban Planning: A 30-Year Journey–A Review. Proceedings of the Institution of Civil Engineers-Urban Design and Planning. Epub ahead of print 2025. https://doi.org/10.1680/jurdp.25.00009

Sharifi A, Khavarian-Garmsir AR (2020) The COVID-19 pandemic: impacts on cities and major lessons for urban planning, design, and management. Sci Total Environ 749:142391

Shishegar N, Boubekri M (2022) Lighting up living spaces to improve mood and cognitive performance in older adults. J Environ Psychol 82:101845

Sobhaninia S, Amirzadeh M, Lauria M et al (2023) The relationship between place identity and community resilience: evidence from local communities in Isfahan, Iran. Int J Disaster Risk Reduct 90:103675

Tabrizi N, Lak A, Moussavi ASMR (2023) Green space and the health of the older adult during pandemics: a narrative review on the experience of COVID-19. Front Public Health 11

Taheri S, Ghasemi M, Shabani A (2021a) Compilation of therapeutic gardens design guidelines with emphasis on promoting the health of the elderly with alzheimer's-systematic review. Haft Hesar J Environ Stud 10(36):117–134
Taheri S, Ghasemi M, Shabani A (2021b) Evaluating the literature of therapeutic landscapes with an emphasis on the search for the dimensions of health: a systematic review. Soc Sci Med 275:113820. https://doi.org/10.1016/j.socscimed.2021.113820
Taheri S, Ghasemi M, Shabani AH (2023) Physical dimension analysis in therapeutic landscape of Shahid Rajaei Park in Isfahan. Geogr Urban Space Dev 10(1(20)):89–114
Taheri S, Khalilpour F, Ashayeri M et al (2024) Handicrafts as cultural creative clusters: a spatial-cultural planning approach for the regeneration of the urban historical fabrics. Int J Tour Cities 10(4):1393–1413
Taheri S, Shabani A, Ghasemi Sichani M (2019) The Role of Therapeutic Landscape in Improving Mental Health of People with PTSD. In: Psychological Trauma. IntechOpen
Taheri S, Shabani A, Istrate A-L et al (2025a) Step up! promoting physical activity in aging populations: a machine learning analysis of park characteristics. Cities Health. Epub ahead of print 2025. https://doi.org/10.1080/23748834.2025.2586869
Taheri S, Shabani A, Istrate A et al (2025b) Sense of belonging and attachment in urban green spaces: a qualitative study of older adults in Isfahan, Iran. Urban Studies. Epub ahead of print 2025. https://doi.org/10.1177/00420980251396546
Tang S, Zhou J, Lin S et al (2022) Where is my home? Sense of home among rural migrant women in contemporary China. Geoforum 129:131–140
Torku A, Chan APC, Yung EHK et al (2022) Detecting stressful older adults-environment interactions to improve neighbourhood mobility: a multimodal physiological sensing, machine learning, and risk hotspot analysis-based approach. Build Environ 224:109533
Tucker I (2010) Everyday spaces of mental distress: the spatial habituation of home. Environ Plan D: Soc Space 28(3):526–538
Verderber S, Koyabashi U, Dela CC et al (2023) Residential environments for older persons: a comprehensive literature review (2005–2022). Health Environ Res & Des J 16(3):291–337
Vitale V, Martin L, White MP et al (2022) Mechanisms underlying childhood exposure to blue spaces and adult subjective Well-being: an 18-country analysis. J Environ Psychol 84:101876
Weinberger AB, Garside EW, Christensen AP et al (2022) Effects of expertise on psychological responses to buildings and natural landscapes. J Environ Psychol 84:101903
Wu H, Ji R, Jin H (2023) Parental factors affecting children's nature connectedness. J Environ Psychol 87:101977
Xu T, Nordin NA, Aini AM (2022) Urban green space and subjective Well-being of older people: a systematic literature review. Int J Environ Res Public Health 19(21):14227
Yang X, Fan Y, Xia D et al (2023) Elderly residents' uses of and preferences for community outdoor spaces during heat periods. Sustainability 15(14):11264
Yu J, Zhang L, Shi Q (2024) Impact of green space on older adults' quality of life based on spatial and non-spatial analyses. Land 13(11):18746
Zhang Y, van Dijk T, Tang J, et al. (2015) Green space attachment and health: A comparative study in two urban neighborhoods. Int J Environ Res Public Health 12(11). MDPI AG: 14342–14363
Zhang Z, Feng L, Hu L et al (2022) Older residents' sense of home and homemaking in rural-urban resettlement: a case study of "moving-merging" community in Shanghai. Habitat Int 126:102616

From Nurtured Cities to Nature-Led Futures: Reflections, Lessons, and Pathways Ahead

Ali Cheshmehzangi, Sara Alidoust, Wendy Y. Chen, and Richard Fuller

Abstract This concluding chapter synthesizes the central arguments of *Nature in Cities, Nurturing Cities*, reframing the relationship between nature and urbanism as a structural, rather than supplementary, dimension of city-making. The book illustrates how nature is essential to ecological stability, social cohesion, cultural identity, and physical and mental health through three thematic pillars: human wellbeing, climate resilience, and social belonging. It makes the case for a paradigm change away from discrete interventions like parks and cooling techniques toward systemic initiatives that integrate natural processes into the very metabolism of urban systems.

Looking ahead, the chapter positions nature-based urbanism as the next frontier: a paradigm in which design, planning, governance, and policy are all influenced by nature as essential infrastructure. It ends by urging academics, professionals, and decision-makers to continue this agenda and imagining cities as adaptive living systems, where ecological and human life co-evolve. This shifts the focus from cities that only host nature to cities that develop from it, establishing the theoretical groundwork for more resilient, just, and regenerative urban futures.

Keywords Urban sustainability · Ecology · Resilient cities · Public health · Sustainable urbanism · Placemaking · Urban futures · Nature · Nature-based urbanism · Nurturing cities

---

A. Cheshmehzangi (✉) · S. Alidoust
School of Architecture, Design and Planning, The University of Queensland, Brisbane, Australia
e-mail: a.chesh@uq.edu.au

A. Cheshmehzangi
School of Architecture and Urban Planning, Hunan University, Changsha, China

College of Architecture and Urban Planning (CAUP), Tongji University, Shanghai, China

W. Y. Chen
Department of Geography, The University of Hong Kong, Hong Kong, Hong Kong

R. Fuller
School of the Environment, The University of Queensland, Brisbane, Australia

Centre for Biodiversity and Conservation Science, The University of Queensland, Brisbane, Australia

A. Cheshmehzangi et al. (eds.), *Nature in Cities, Nurturing Cities*, Urban Sustainability,
https://doi.org/10.1007/978-981-95-5487-4_12

1 Reflecting on the Journey: Nature's Role in Urban Futures

For decades, if not for centuries, cities have been constructed to dominate their surroundings rather than coexist with them; this chronicle of urban history has been written in concrete, steel, and asphalt. Economic output, spatial control, and engineering efficiency dominated urban planning for over a century. Parks were viewed as recreational afterthoughts, boulevards lined with trees as aesthetic flourishes, and rivers as impediments that needed to be buried or channelled. Even though this strategy appeared to be effective in promoting urbanization and industrial growth, it has now shown its limitations. It has resulted in cities that overstress people and the environment by overheating and overconsuming. A fundamental reassessment of what cities are and how they operate is now required due to the climate catastrophe, public health crises, and growing social injustices. In response to that requirement, this book reinterprets nature as a fundamental component of urban life, crucial to the sustainability, adaptation, and evolution of cities, rather than as a passive backdrop.

This intellectual and practical shift from concrete-centric urbanism to ecology-centric futures is traced in the chapters included here. They prove, with detail and evidence, that nature is not an extravagance that should be introduced after the "real" labour of creating cities is over. Instead, it serves as the framework for achieving cohesiveness, resilience, and sustainability in urban systems. There is more to this change from nature in cities to nature as cities than meets the eye. It signals a paradigm shift in which natural processes play a key role in how humans plan, manage, and live in urban areas rather from being merely incidental.

The book's three-part structure reflects this evolving logic. The first section, on *Nature and Human Wellbeing*, shows how closely ecological existence and human situation are intertwined. There is no question about the evidence: being close to nature lowers stress, improves mental health, increases physical fitness, and fortifies social ties. These advantages are essential rather than incidental. They are the result of social dynamics, psychological processes, and biological mechanisms that have developed over thousands of years. However, these ties run the risk of being broken as urbanization picks up speed, creating what some scholars refer to as a "public health deficit of disconnection". This section's chapters address that reality and demonstrate how planning interventions, policies, and community projects can help to rebuild these crucial connections. They support a health agenda that acknowledges that parks, urban forests, and waterways are also public health engines and places green infrastructure on par with hospitals and transportation systems.

The second part, *Climate Resilience and Urban Cooling*, situates nature as a strategic response to planetary pressures. Cities are increasingly at the forefront of climate change's effects, from more frequent flooding to heatwaves that get worse. These issues cannot be resolved by traditional grey infrastructure, such as power lines, pavements, and pipelines. Green infrastructure, by contrast, offers multifunctional benefits: it moderates temperatures, absorbs stormwater, stores carbon, and buffers biodiversity. Case studies from Australia, China, and Iran show how integrated architecture, blue-green networks, and diverse vegetation may change microclimates and lower urban heat. Importantly, these tactics show that adapting to climate change is an ecological issue as well as a technical one, requiring that urban form be in harmony with the cycles and capabilities of natural systems.

The final part, *Social Belonging and Placemaking*, explores the cultural and emotional dimensions of nature in the city. In addition to being natural resources, parks and greenways serve as places of social interaction, identity, and remembrance. They offer areas where communities congregate, where links between generations are forged, and where a feeling of community is fostered. These chapters demonstrate how environment supports social resilience and cultural continuity in a variety of settings, such as Pacific Island cultures readjusting to disasters and senior citizens finding "home" in urban parks. Nature-based placemaking serves as a platform for justice, inclusivity, and community empowerment in addition to being aesthetically pleasing and recreational.

These three parts/themes come together to create a coherent narrative: cities cannot prosper without nature since it supports the essential attributes that enable urban existence, such as resilience, identity, safety, and health. Thus, the change mapped in this work is structural rather than incremental. It pushes policymakers, planners, designers, and residents to view nature as the main infrastructure of the urban century rather than as a competing land use.

This reinterpretation (or reframing) is a political endeavour as well as a scientific need. It necessitates breaking down long-held beliefs on land values, development goals, and the purported trade-off between green and growth. It advocates for design that supports natural processes rather than interferes with them, integrative rather than segregated governance methods, and a revitalized cultural perception of cities as a part of their ecosystems rather than as something separate from them. If the industrial city of the past century was a triumph of human engineering, the city of the future must be a triumph of ecological intelligence—a place where the pulse of urban life and the dynamics of the living world move in synchrony.

This book's journey—across wellbeing, resilience, and belonging—is a step toward that future. It demonstrates that preserving nature in urban areas requires reinventing their core rather than just softening their boundaries. In doing so, it presents a convincing argument: cities that build their foundations on living systems that support all of us rather than concrete will thrive in the ensuing decades.

2 Key Insights and Takeaways: What We've Learned

As evidenced by the first theme pillar, Nature and Human Wellbeing, nature is not only advantageous to urban living but also essential to human biology and psychology. Exposure to green spaces, tree canopy, water bodies, and biodiverse habitats has been shown to reduce stress hormones, improve cognitive performance, and lessen symptoms of anxiety and depression. This is supported by evidence presented in each chapter. These benefits are not fleeting; they show up as better cardiovascular health, immune system bolstering, and physiological balance. The social aspects of nature, such as chances for communication, social cohesiveness, and collective identity, are just as transformative beyond personal health. Cities that incorporate nature into their everyday operations are more resilient, healthier, and connected. However, the data also reveals significant disparities in access: populations with lower socioeconomic status frequently have the least amount of exposure to green spaces. Deliberate policy interventions, cross-sector collaboration, and a reinterpretation of health infrastructure to incorporate natural systems as essential public assets are all necessary to address these inequities.

The strategic role of nature in tackling global issues that jeopardize the sustainability of cities is highlighted by the second pillar, Climate Resilience and Urban Cooling. In the face of increasing extreme heat, heavy rainfall, and flooding, traditional grey infrastructure cannot keep up. Green infrastructure, multilayered vegetation, and blue-green networks are examples of nature-based solutions that are effective at reducing energy consumption, improving air quality, absorbing stormwater, and regulating temperatures. Empirical research shows that vegetation complexity and water retention are critical variables shaping microclimates, while the spatial distribution of natural elements significantly influences outcomes like urban heat island intensity. These results also highlight an important planning realization: context determines effectiveness. If airflow is restricted, vegetation that cools a suburban park may trap heat in a dense city centre. Therefore, a site-specific, nuanced strategy, and governance structures that can incorporate ecological knowledge into urban design are necessary for successful adaptation. Climate resilience underpinned by urban nature is an ecological rethink of urban structure rather than a technical improvement.

The third and final pillar, Social Belonging and Placemaking, extends the argument beyond health and adaptation to the realm of identity and collective meaning. In divided urban cultures and societies, nature-based areas play an increasingly important role as anchors for social interaction, emotional connection, and cultural memory. These case studies demonstrate how nature influences the social fabric of cities, whether it is through the reconstruction of a Pacific island community's sense of place after a disaster, the creation of social bonds by young people in biophilic parks, or the discovery of continuity and a sense of belonging by elderly people in urban gardens. Crucially, this factor emphasizes the effectiveness of participatory design and co-creation. Involving communities in the process of creating the places they live in ensures stewardship, equity, and cultural relevance.

Across all three pillars, common threads emerge: nature offers multifaceted advantages when it is incorporated into local contexts, co-produced with people, and integrated into governance structures. Together, these insights redefine nature not as a complementary amenity but as the connective tissue that binds urban health, climate action, and social resilience into a coherent whole.

3 Beyond this Book: Toward Nature-Based Urbanism

The insights collected in this book suggest that the way cities are planned and constructed needs to change. Specifically, "nature-based urbanism" (a topic that is covered in the second volume of this trilogy) should replace isolated greening initiatives as the framework for urban growth in our time. These interventions, which range from healing parks to corridors that link to cooling vegetation and buffering waterscapes, are important but only the beginning. Future cities must adopt systemic concepts where ecological processes serve as the foundation for the planning, management, and control of the urban environment, moving beyond such isolated projects. This entails considering cities as dynamic metabolic systems in which resources, energy, water, and biodiversity flow through interdependent networks.

Nature must no longer be viewed as an adjunct to infrastructure, but rather as the infrastructure itself, influencing zoning, investment priorities, property valuation, and spatial planning. Cities can only genuinely provide resilience, equity, and sustainability at scale when nature is ingrained in finance and policy institutions as a fundamental logic rather than an optional feature.

This transformation also opens new frontiers of innovation and integration. Ecological design is simply one aspect of nature-based urbanism's future; other aspects include its confluence with digital intelligence, models of the circular economy, and active governance. Predictive analytics, sensor networks, and urban digital twins can boost ecological performance, and nature-based business models and circular resource flows can reorient economic value to emphasize regeneration over extraction. Social innovation will ground these changes in shared responsibility and lived experience, from green job creation to community stewardship models.

This foundation will be expanded upon in the upcoming two volumes, "*Nature-Based Urbanism*" and "*Designing Nature into Cities*", which will convert ideas into tactics for structural change. They will demonstrate how cities might transform from habitats for bits of nature to dynamic, regenerative systems where ecological and urban life are intertwined—a model for both planetary survival and sustainable urbanism.

4 Concluding Reflections: Cities as Living Systems

The story of cities has always been a story of human ambition, i.e., to build, to connect, to endure. Yet in our pursuit of progress, we often forgot the foundation upon which every thriving settlement has been built: the living systems that sustain us. Through examples and supporting data, this book has demonstrated that nature is the lifeblood of urban life rather than its backdrop; hence the title of this volume is focused on "nature in cities, nurturing cities". Adopting this reality makes cities healthier, more equitable, and more resilient in addition to being greener. They are places where people feel like they belong, where innovation coexists with biodiversity, and where everyday life is governed by the cycles of the environment. In a century characterized by resource scarcity, social fragmentation, and climate unpredictability, acknowledging nature as infrastructure, community, and culture is an existential necessity rather than an aesthetic option.

The path ahead is both a challenge and an invitation. It challenges people, legislators, and planners to consider cities as adaptive ecosystems, or in another words, places that learn, regenerate, and change instead of traditional silos and linear solutions. It calls for economics that reward restoration over extraction, technologies that enhance rather than replace natural intelligence, and government that values soil and canopy as much as steel and concrete. The most transformative cities of the future will be those that do not merely accommodate nature but grow from it—where wetlands determine infrastructure logic, forests define neighbourhood identity, and rivers direct urban form—will be the most revolutionary. This is not a utopian dream; it is an achievable model that is already gaining traction and taking roots in initiatives and regulations all around the globe.

As this volume closes, it does so not with a conclusion but with a call to action. The concepts discussed here are seeds for new research, new forms of governance, and new social contracts with the natural world. Those who see that the city is a living system integrated into the greater fabric of Earth's ecosystem, rather than an artefact distinct from it, will be the ones driving urbanism in the future. Our task now is to nurture that living city, to let it grow in harmony with the planet that makes all urban life possible.